# Air Quality and Health

*Edited by Ayşe Emel Önal*

Published in London, United Kingdom

Air Quality and Health
http://dx.doi.org/10.5772/intechopen.98023
Edited by Ayşe Emel Önal

Contributors
Özkan Çapraz, Ali Deniz, Riccardo Pansini, Lei Shi, Tadesse Weyuma Bulto, Birhanu Chalchisa Werku, Trupti Das, Minati Behera, Jyotishree Nath, Sony Pandey, Ramasamy Boopathy, Odubanjo D. Adedolapo, Ferdi Tanir, Burak Mete



First published in London, United Kingdom, 2022 by IntechOpen
IntechOpen is the global imprint of INTECHOPEN LIMITED, registered in England and Wales, registration number: 11086078, 5 Princes Gate Court, London, SW7 2QJ, United Kingdom

British Library Cataloguing-in-Publication Data
A catalogue record for this book is available from the British Library

Additional hard and PDF copies can be obtained from orders@intechopen.com

Air Quality and Health
Edited by Ayşe Emel Önal
p. cm.
Print ISBN 978-1-80355-693-2
Online ISBN 978-1-80355-694-9
eBook (PDF) ISBN 978-1-80355-695-6

# Meet the editor

Ayşe Emel Önal is a medical doctor, public health specialist, and professor. She graduated from the Faculty of Medicine, Istanbul University in 1985. Since 2016, she has been working as the head of the Department of Public Health at the same university. She has edited and authored many national and international books and book chapters. She has many research articles in national and international journals and papers in national and international congresses to her credit. Dr. Önal has supervised numerous graduate students in occupational health and safety and public health. Her main research areas are epidemiology and environmental health.

# Contents

# Preface

Air pollution is one of the most important causes of morbidity and mortality, as it causes cardiovascular and respiratory system diseases, which are the most common causes of death today. According to the World Health Organization (WHO), fine particulate matter is 2.5 microns or less in diameter and is taken into the respiratory system by inhalation where it causes mainly chronic bronchitis and chronic obstructive pulmonary diseases. In addition, particulate substances with a diameter of less than 0.5 microns pass through the alveoli into the bloodstream and cause cardiovascular diseases. It is known that ozone in the atmosphere has a protective role against ultraviolet rays, but the presence of high concentrations of ozone on the earth, together with sulfur dioxide and nitrogen oxides, adversely affects the respiratory and cardiovascular systems. In addition, sulfur dioxide, nitrogen oxide, Volatile Organic Compounds (VOCs), and polycyclic aromatic hydrocarbons (PAHs), which are the most important pollutants in indoor air, can cause reproductive and central nervous system disorders and various cancers. Inhalation of high levels of carbon monoxide can cause acute poisoning and death. Lead is a heavy metal and is another one of the main outdoor air pollutants. While chronic lead exposure can lead to diseases such as anemia, memory impairment, and hypertension, lead intoxication may develop in acute work-related lead exposures in the indoor environment. Air pollution is one of the most important components of environmental pollution and therefore it is one of the most important causes of climate change and natural disasters. It also facilitates the spread of many infectious diseases directly and indirectly. Biological pollution of the air leads to respiratory tract infections. Dust facilitates the spread of biological pollution.

Rapid population growth, rapid industrialization, excessive use of motor vehicles, and the use of fossil fuels for heating are the most important causes of outdoor air pollution. The use of biomass, wood and other solid fuels for heating in the indoor environment and the use of cigarettes and tobacco are the most important indoor air pollutants. Air quality standards and guidelines for different air pollutants are used for air quality management.

This book highlights the importance of clean air for public health, draws attention to the sources of air pollution, and proposes some solutions and interventions to prevent air pollution. I would like to thank the chapter authors for their valuable contributions. It is my wish that efforts for clean air will continue.

**Ayşe Emel Önal**
Istanbul Medical Faculty,
Department of Public Health,
Istanbul University,
Istanbul, Turkey

Section 1

# Air Quality and Health Relationship

Chapter 1

# Air Quality and Health in Ethiopia

*Tadesse Weyuma Bulto and Birhanu Chalchisa Werku*

## Abstract

In Ethiopia, people who live in cities with rapid urbanization and high industrialization suffer greatly from air pollutants. These air pollutants, which include particulate matter, sulfur dioxide ($So_2$), nitrogen oxide ($No_x$), carbon monoxide (CO), and ozone ($O_3$), often exceed National Ambient Air Quality Standards. When we compared African regions, we found that the population of Eastern and Western Sub-Saharan Africa is greatly affected by air pollution, whereas the population of Southern and Central Sub-Saharan African provinces is less affected by air pollution, specifically through particulate matter. Air pollution is a major issue for both developing and developed countries. The main sources of air pollution in Ethiopia are open burning of refuse, vehicular emissions, and traditional practice. Among these, open burning causes the greatest air pollution in metropolitan cities like Addis Ababa. This chapter examines the effect of air pollution on human health in Ethiopia. In high concentrations, air pollutants can harm both human and ecosystem health. The effect of air pollution on human health includes irritation of the eyes, coughing and breathing difficulties, worsening of existing lung and heart problems, asthma, and increased risk of heart attack. Air quality and health human are directly related. As such, air quality management plans for developed and developing countries are needed to minimize air pollution and reduce risk to human health.

**Keywords:** air quality, health, air pollution, particulate matter

## 1. Introduction

Air pollution is the largest single environmental risk to health, responsible for an estimated seven million premature deaths globally every year. Air pollution is an important determinant of health [1]. Air pollution presents a global problem that undermines health and economic productivity. Data from the World Health Organization (WHO) shows that nine out of ten people breathe air containing high levels of pollutants, with low- and middle-income countries (LMICs) bearing the brunt of poor air quality [2]. Principally, open burning of refuse is a major source of high air pollution in metropolitan cities. This chapter identifies pollutant concentration of particulate matter ($PM_{2.5}$) emission. $PM_{2.5}$ is harmful to people who are unusually sensitive to particulate pollution and have health problems [3]. In addition, vehicle-related particulate matter is a main source of air pollution in the urban environment [4].

Particulate matter contains both solid particles and liquid droplets found in the air's atmosphere. There are artificial (humanmade) and natural sources of air

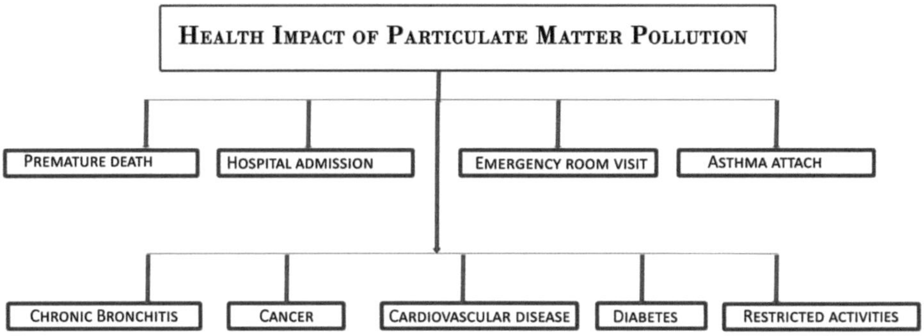


**Figure 1.**
*Effect of fine particulate matter on human health.*

pollution. Sources of fine particles include motor vehicles, power plants, wood burning, and industrial processes. Sources of coarse particles include crushing and dust from paved or unpaved roads [5]. According to different studies, the main sources of primary fine particles are cars and trucks (diesel engines), open burning, wildfires, woodstoves, outdoor wood boilers, cooking, dust from roads and construction, agricultural operations, and coal and oil-burning boilers. The main sources of secondary fine particles are power plants and industrial processes such as oil refining and pulp and paper production.

Knowledge of the effects of air pollutants on human health is a prerequisite for the development of effective policies to reduce the adverse impact of ambient air pollution [6]. Due to the health effects of air pollutants, billions of people and animals die yearly. Fine particulate matter ($PM_{2.5}$) has been determined to be a serious worldwide air problem during the last several decades. People who live in cities with rapid urbanization and high industrialization suffer more from air pollutants than those individuals living in rural areas. In addition, particulate matter pollution contributes the most to global burden of disease in developing countries. Air pollution can lead to increased risk of lung cancer, heart disease, bronchitis, and other cardiorespiratory conditions (**Figure 1**) [7].

Air quality is a cause for concern in Ethiopia, particularly in cities, and air pollutants including particulate matter, sulfur dioxide (SO2), nitrogen oxide (NOx), carbon monoxide (CO), and ozone (O3) often exceed the National Ambient Air Quality Standards (NAAQS). Daily records of raw concentration of air pollutant and air quality index data acquired at one-hour intervals can be found on the AirNow website of Addis Ababa's central monitoring station.

## 2. Effect of air pollution on human and ecosystem health

Air pollution has both short- and long-term effects on human health. Some short-term effects include irritation of the eyes, nose, and throat, and coughing and breathing difficulties. Long-term effects of air pollution can cause cancer and damage to the immune, reproductive, and respiratory systems. In extreme cases, air pollution can even cause death.

A three-year study (2017–2021) of fine particulate pollution in Addis Ababa found that the daily mean (SD) $PM_{2.5}$ concentration was 42.4 $\mu g/m^3$ (15.98). A total of 502 deaths (4.44%) were attributable to current air pollution levels referenced to the 35 μg/m$^3$ WHO interim target annual level and 2043 (17.7%) at the WHO 10 μg/m$^3$ annual guideline [8]. Another study aimed to identify the major sources of $PM_{2.5}$ in Ethiopia's capital city and found that 28% comes from vehicular sources, 18.3% comes from biomass burning, and 17.4% comes from soil dust, all of which comprise about two-thirds of total $PM_{2.5}$ mass, followed by sulfate at 17.4% [9].

Data from a study of particulate matter with aerodynamic diameter < 10 μm ($PM_{10}$) in Addis Ababa show that maximum emission occurs around 7:00 am daily with secondary peaks in the late afternoon and evening. This suggests that $PM_{10}$ is emitted during periods associated with motor-vehicle traffic, food preparation, and heating of homes. The morning concentration is likely accentuated by stable atmospheric conditions associated with overnight surface temperature inversions [10]. Air contamination influences the human health and environmental wellbeing of the ecosystem. Particulate matter is a series of issues from major air pollutants in the atmosphere [11].

When we compared African regions, we found that the population of Eastern and Western Sub-Saharan Africa is greatly affected by air pollution, whereas the population of the Southern and Central Sub-Saharan African provinces is less affected by air pollution, specifically through particulate matter. Priority should be given to air quality management to improve human and environmental health and reduce the global burden of disease of African provinces [11].

Health risk assessments of air pollution are being developed for a variety of policy scenarios, using different methodologies and spatial and temporal scales [12]. Air pollution is a serious health and environmental problem. Poor air quality has been linked to numerous diseases and is a significant public health issue related to urban planning [13]. Ethiopians face quality of life and livelihood challenges associated with sub-optimal sanitation, dependence on biomass energy, and decreasing agricultural productivity [14]. Detailed knowledge of the effects of air pollutants on human health is a prerequisite for the development of effective policies to reduce the adverse impact of ambient air pollution (**Figure 2**) [15].

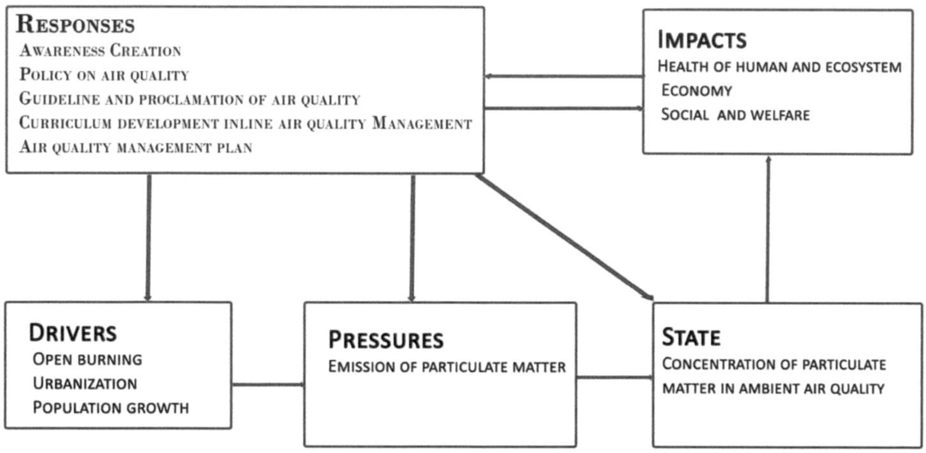


**Figure 2.**
*DPSIR diagram for particulate matter pollution.*

## 3. Conclusion

This chapter discussed air pollution and its effects on human health in Addis Ababa, Ethiopia. It presented data from scientific studies showing the types and amounts of fine particulate matter pollution present in the African city. Air pollution is a transboundary problem that particularly affects vulnerable groups. Many Ethiopian cities face challenges to quality of life and livelihood associated with sub-optimal sanitation, dependency on biomass energy, and decreasing agricultural productivity. As such, the country, as well as other developing and developed countries worldwide, requires an air quality management plan to minimize pollution and mitigate its negative effects on humans and the environment.

## A. African provinces exposed to air pollution, total avoided deaths, percent of baseline mortality, deaths per 100,000 population, economic benefits, and avoided life years lost.

| S.N | Region and Country | Population Affected | Avoided Deaths (Total) | 95% CI | % of Baseline Mortality | Deaths per 100,000 | Avoided Deaths (% Population) | Economic Benefits (2011 US) | Avoided Life Years Lost |
|---|---|---|---|---|---|---|---|---|---|
| 1 | Central Sub-Saharan Africa | 40,000,000 | 9800 | 7700–12,000 | 5.6830 | 24.62 | 0.0246 | 4,100,000,000 | 9500 |
| 2 | Eastern Sub-Saharan Africa | 150,000,000 | 28,000 | 22,000–34,000 | 4.2522 | 18.53 | 0.0185 | 3,400,000,000 | 28,000 |
| 3 | Southern Sub-Saharan Africa | 37,000,000 | 7700 | 6100–9400 | 5.0706 | 20.69 | 0.0207 | 6,500,000,000 | 9400 |
| 4 | Western Sub-Saharan Africa | 140,000,000 | 26,000 | 21,000–32,000 | 4.2948 | 18.22 | 0.0182 | 5,300,000,000 | 21,000 |

## Author details

Tadesse Weyuma Bulto[1*] and Birhanu Chalchisa Werku[2]

1 Department of Environmental Management, Faculty of Urban Development Studies, Kotebe Metropolitan University, Addis Ababa, Ethiopia

2 Department of Rural Development and Agricultural Extension, Shambu Campus, Wollega University, Wollega, Ethiopia

*Address all correspondence to: tadesseweyuma@kmu.edu.et

## References

[1] WHO. Air quality and health. 2018.

[2] WBG. Safe to Breathe? Analyses and Recommendations for Improving Ambient Air Quality Management in Ethiopia. 2021

[3] Bulto TW. "Impact of open burning refuse on air quality: In the case of 'Hidar Sitaten' at Addis Ababa, Ethiopia," Environmental Health Insights. Jan 2020; **14**:1-11. DOI: 10.1177/1178630220943204

[4] Jida SN, Hetet J, Chesse P, Guadie A. Roadside vehicle particulate matter concentration estimation using artificial neural network model in Addis Ababa , Ethiopia roadside vehicle particulate matter concentration estimation using artificial neural network model in Addis Ababa, Ethiopia. Journal of Environmental Sciences. 2020;**101**(September):428-439. DOI: 10.1016/j.jes.2020.08.018

[5] EPA. Air Quality Index Air Quality and Air Quality Index A Guide to Air Quality and Your Health. 2000. p. 10

[6] WHO. 20060929_factsheet.pdf.

[7] Avis W, Weldetinsae A, Singh A, and Getaneh Z. Vulnerability scoping study: Air pollution exposure and public transport in Addis Ababa. 2020

[8] Kumiea A et al. Fine particulate pollution concentration in Addis Ababa exceeds the WHO guideline value. Environmental Epidemiology. 2020;**5**:1-9. DOI: 10.1097/EE9.0000000000000155

[9] Tefera W et al. Source apportionment of fine organic particulate matter (PM 2.5) in Central Addis Ababa, Ethiopia. 2021

[10] Etyemezian V et al. Results from a pilot-scale air quality study in Addis Ababa, Ethiopia. Atmospheric Environment. 2005;**39**(40):7849-7860. DOI: 10.1016/j.atmosenv.2005.08.033

[11] Bulto TW. Influence of particulate matter on human health in selected African provinces: Mini-review. Reviews on Environmental Health. 2020. DOI: 10.1515/reveh-2020-0013

[12] WHO. WHO Expert Meeting: Methods and Tools for Assessing the Health Risks of Air Pollution at Local, National and International Level. 2014

[13] Yan Y. Air pollution in China: A study of public perception. 2016.

[14] Abadi N, Gebrehiwot K, Techane A, Nerea H. Links between biogas technology adoption and health status of households in rural Tigray, northern Ethiopia. Energy Policy. 2017;**101**(November 2016):284-292. DOI: 10.1016/j.enpol.2016.11.015

[15] WHO. Health Aspects of Air Pollution with Particulate Matter, Ozone and Nitrogen Dioxide. 2003

Chapter 2

# Air Quality and Health in West Africa

*Odubanjo D. Adedolapo*

## Abstract

One of the most important elements for survival is air. Its significance cannot be overstated, necessitating proactive measures and regulations to ensure clean air in our atmosphere. Africa is one of the continents with the worst air quality. According to NASA modelling research, air pollution causes approximately 780,000 premature deaths per year in Africa. Experiments were carried out by the European-African consortium DACCIWA to investigate the causes and effects of air pollution by looking at the entire chain of natural and human-made emissions, from formation to dispersion to repercussions. The findings suggest that air pollution has already reached a dangerous threshold for human health in most West African countries. The aim of this chapter is to highlight and increase awareness about the severe risk that air pollution poses to the health of inhabitants of West African countries.

**Keywords:** air pollution, air quality, health, particulate matter, emission, disease, pollutants

## 1. Introduction

Africa is undergoing a lot of major changes. In this century, the continent's population is expected to more than triple, from 13 billion in 2020 to 43 billion in 2100. Cities are rapidly increasing, economies are booming, and life expectancy has nearly doubled. At the same time, as a result of increased fossil fuel burning, ambient air pollution is increasing, and mortality from ambient air pollution has increased from 2613 per 100,000 population in 1990 to 2915 per 100,000 population in 2019. Despite its small volume, this rise is unmatched in history. The most significant gains are found in Africa's most developed countries [1]. African cities have become a substantial source of pollution as a result of rising populations, urbanisation, and resource-intensive industries. African urban growth rates are currently and will likely remain the highest in the world, averaging 3.1–3.8% each year. The World Health Organisation (WHO) estimates that the annual median concentration of PM2.5 in more than half of Africa exceeded 26 $g/m^3$, much surpassing the WHO-set guideline of 10 $g/m^3$ as the annual average for healthy outdoor air. Air pollution monitoring is severely poor; only six of the 47 countries that make up Sub-Saharan Africa can provide long-term data on airborne particulate matter (PM), spanning 16 cities [2]. The automobile pool is the most significant source of pollution in urban areas. Increased traffic congestion is caused by an increase in the number of vehicles on the road and a lack of urban planning, which leads not only to increased air pollution but also to major economic losses in terms of time and fuel. In Sub-Saharan Africa, PM from road traffic is substantially higher than in developed countries. PM2.5 values varied between 40 and 260 g/m$^3$ in a review of eight studies of outdoor air pollution in African cities (covering seven nations), compared to an annual average of 13 g/m$^3$ in urban Europe and 9 g/m$^3$ in urban America in 2019. In four West African cities, road traffic was the leading source of black carbon and PM2.5 (88%), with diesel exhaust being the most significant contributor to PM at roadside in Addis Ababa, Ethiopia, contained several substances linked to negative health impacts, including chromium, cadmium, zinc, and lead [2].

It is now well known that air pollution causes a significant amount of disease. The International Agency for Research on Cancer (IARC) of the World Health Organisation recently has classified outdoor air pollution as carcinogenic to humans, placing it in the same category as cigarette smoke, UV radiation, and plutonium. In 2010, ambient fine particles were responsible for approximately 223,000 lung cancer deaths, with more than half of those fatalities occurring in China and other East Asian countries (Straif et al. 2013). In 2019, an estimated 1.1 million people died in Africa as a result of air pollution (95% UI 932,000–1.26 million). HAP caused an estimated 697,000 fatalities (95% UI 526,000–879,000), ambient PM2.5 pollution caused 383,000 deaths (95% UI 289,000–491,000), and ambient ozone pollution caused 11,300 deaths (95% UI 4800–18,300) [1]. Air pollution is Africa's second leading cause of death. It caused more deaths than tobacco, alcohol, car accidents, and substance abuse. Only AIDS is responsible for more deaths. Chronic lung infections (336,460 deaths; UI: 251,827–430,493), myocardial infarction (223,930 deaths; UI: 185,558–268,252), new-born disorders (186,541 deaths; UI: 152,569–229,402), chronic obstructive pulmonary disease (COPD) (70,479 deaths; UI: 53,765–87,251), and stroke are amongst (193,936 deaths, UI, 165,936–227,196) [1]. Due to the adverse effects of air pollution, clean air guidelines and policies have been approved or backed by the WHO and the United Nations' Sustainable Development Goals. This has aided many countries throughout the world in developing efficient policy, tools and data to control, and even reduce, harmful emissions in cities, hence reducing health burdens. Governments in other regional contexts, on the other hand, continue to deal with rising levels of urban air pollution and underperforming air quality surveillance for a variety of reasons. A prime example is Africa [3].

## 2. Air quality and health in west africa

Pollutant concentrations differ considerably based on a country's population, industrialisation, urbanisation, and economic status, as well as its physical region, emission sources, and meteorology. Air quality research has traditionally focused on North America and Europe's industrialised regions. However, with significant breakthroughs in our understanding, the focus has shifted to quickly emerging Asian regions (particularly China and India) over the last decade. Despite its enormous and quickly rising population, Africa is still a little-studied region when it comes to air pollution [4]. Pollutant concentrations in West Africa are caused by a number of different sources. Anthropogenic sources, such as transportation, industries, and refuse combustion, as well as natural biogenic emissions from plants, fall under this category. The location of the population is one of the most important elements impacting the locations and sources of emissions in West Africa. West Africa's population density is higher around the shore because extensive port infrastructures are present in major coastal towns. Shipping emissions are a major source of pollution along the coast, particularly along major shipping lanes and at offshore oil and gas operations. Within cities, pollution is caused by emissions from industries and power plants, as well as emissions from motorised vehicles and home sources, such as wood-fired stoves for cooking and refuse to burn [4]. Man-made emission sources, such as transportation, household burning, and industries tend to be located adjacent to places with high population density, making anthropogenic emissions one of the most important factors affecting human health. As a result, anthropogenic emissions can be expected to scale as a function of population. As a result of Africa's widespread economic growth, quick rates of urbanisation and industrial development, and growing population, anthropogenic pollution emissions are expected to climb dramatically during the next century. Much of this growth is predicted to take place along the West African coast. While the effects of anthropogenic pollution on human health have been extensively studied around the world, quantifying the negative consequences on the West African population is difficult [4].

The levels of air pollution in ECOWAS cities are rapidly increasing to alarming levels: Nigerian cities, such as Onitsha and Kaduna, are now amongst the world's most polluted, with PM10 concentrations 30 and 21 times higher than the WHO's limit, respectively. To complicate things, all ECOWAS countries are classed as low- or lower-middle-income, implying that their public health systems are vulnerable and that their populations are more susceptible to poverty-related diseases, such as tuberculosis. Long-term exposure to ambient air pollution has been demonstrated to raise the chance of getting active and even drug-resistant tuberculosis infections, putting ECOWAS city dwellers at even greater risk [3]. Despite the great danger that air pollution poses to ECOWAS countries, the availability of air quality data from government sources is a problem. This contributes to the fact that the issue of urban air pollution receives little public attention and is not high on government agendas. Due to a lack of data on air quality, many national and local governments only have a fragmentary understanding of emissions sources, concentrations, and trends. It also means that the efficiency of some governments' forecasting models may be hampered, as they rely on reliable ground-level readings [3]. ECOWAS countries trail behind other African countries in terms of official air quality monitoring and related health expertise, such as South Africa, which has studies employing air quality monitoring data to identify health hazards. There has been no comprehensive review of available knowledge on this subject to date, as the few existing reviews on air quality and health in the Sub-Saharan African region (none of which were specific to West Africa) have focused on indoor air pollution or specific populations, such as children, and have not assessed the status of air quality monitoring or policy [3].

Accra (Ghana) is one of the ECOWAS cities that has achieved significant improvements in air quality monitoring recently. The Ghana Environmental Protection Agency has established an Air Quality Management Plan that covers up to 10 districts in Accra and its environs, in collaboration with the US Environmental Protection Agency, the US Agency for International Development (USAID), and UNEP [3]. In the case of Nigeria, [5] indicate that the monitoring network is "scantily disseminated," implying that there are stations, but no mention of their location or functionality. The Nigerian Meteorological Agency established five stations across the country in the early 2000s, however it is speculated that they have not been operational since 2013, probably due to a lack of technical expertise required to properly maintain equipment. Furthermore, air quality monitoring equipment is occasionally damaged, possibly as a result of a lack of public knowledge of pollution-related health problems [3]. [6] particularly highlight the six measuring sites in Dakar, Senegal, that are now operational. The city's Air Quality Management Centre (CGQA—Centre de Gestion de la Qualité de l'Air) manages these sites in the Guédiaway, Medina, Yoff, Bel-air, HLM, and Cathedral districts. Another study conducted by [7] in Dakar combined air quality monitoring data, Ministry of Health data, models, and seasonal variability of Saharan dust influence to look at patterns of respiratory diseases, such as asthma, bronchitis, and tuberculosis. They discovered a significantly greater incidence of respiratory conditions in Dakar than in other parts of the country, implying that anthropogenic air pollution has a significant impact.

In summary, only two of the 15 ECOWAS countries appear to have operational government air quality monitoring stations. Cabo Verde, Gambia, Guinea, Guinea-Bissau, Liberia, and Sierra Leone had no information [3].

### 2.1 The DACCIWA project: dynamics-aerosol-chemistry-cloud interactions in west africa

It is clear that there is a major shortage of knowledge and observational data in West Africa regarding atmospheric composition and air pollution. The Dynamics-Aerosol-Chemistry-Cloud Interactions in West Africa (DACCIWA) project, funded by the European Commission's Framework 7 programme with €8.75 million, has helped to close this knowledge gap [8]. The partnership, which includes universities, research institutes, and operational meteorological and climate services, is made up of 16 partners from four European and two West African nations. The DACCIWA consortium's expertise spans atmospheric chemistry, aerosol science, air pollution, and their consequences for human and ecological health, as well as atmospheric dynamics, climate science, cloud microphysics, and radiation. It has competence in the ground, air, and space observations, as well as modelling and impact studies [8]. DACCIWA intends to contribute to 10 fundamental objectives, the first nine of which are research-oriented. The objectives are as follows [8]:

- During the wet season in southern West Africa, determine the impact of different emission sites (natural and anthropogenic), as well as mixing and transportation processes, on aerosol composition.

- Evaluate the effects of surface pollutants (particularly particulate matter and ozone) on human health, ecological health, and agricultural output.

- Accurately measure two-way aerosol-cloud interaction with a focus on cloud condensation nuclei properties and distribution, as well as their impact on cloud characteristics and aerosol removal by precipitation.

- Determine the factors that influence low-level cloud formation, persistence, and breakup.

- Determine the meteorological factors that influence precipitation.

- Measure the effect of aerosols and clouds on energy and radiation budgets, with an emphasis on aerosol effects on cloud characteristics.

- Analyse meteorological, chemical, and air quality models, as well as satellite precipitation, radiation, cloud, and aerosol retrieval.

- Examine the impacts of precipitation and cloud radiative forcing on the water budget and the circulation of the West African monsoon.

- Evaluate the socioeconomic consequences of future changes in emissions, climate, and land use on human health, ecosystem health, agricultural output, and water availability.

- Inform the general public, scientists, operational centres, and policymakers about critical results.

To achieve these goals, DACCIWA science is divided into seven scientific Work Packages (WPs) that coincide with the major study areas, which are [8];

1. Boundary Layer Dynamics

2. Air Pollution and Health

3. Atmospheric Chemistry

4. Cloud-Aerosol Interactions

5. Radiative Processes

6. Precipitation Processes

7. Monsoon Processes

The lack of data was a key stumbling block to achieving the DACCIWA research goals outlined above. To address this, DACCIWA conducted major field campaigns in SWA in June and July 2016, which included coordinated flights with three research aircraft—the British Antarctic Survey (BAS) DHC-6 Twin Otter, the Deutsches Zentrum für Luft- und Raumfahrt (DLR) Falcon 20, and the Service des Avions Francais Instrument'es pour la Recherche en Environnement (SAFIRE) ATR-42 and a wide range of surface-based instrumentation in Kumasi, Ghana, Savé, Benin, and Ile-Ife, Nigeria [9]. During the 3-week campaign period (June 29–July 16, 2016), the three aircraft flew 50 research flights across Cote d'Ivoire, Ghana, Togo, and Benin. One of the purposes of the research flights was to collect air pollution measurements all across some of the world's most populous cities. Abidjan.

(Côte d'Ivoire), Accra and Kumasi (Ghana), Lome (Togo), and Cotonou and Save were amongst the cities targeted by the DACCIWA flight tracks (Benin). Each plane carried an identical cargo of instruments for measuring weather patterns, chemical concentrations, and aerosol and cloud characteristics. Most of the measurements were made onboard the plane; however, some of the samples acquired in flight were analysed offline [4].

Notwithstanding the project's primary focus on meteorological conditions, it came to a number of important conclusions about air quality and its effects on human health. In the cities of Abidjan and Cotonou, measurements of tiny particles suspended in the air (known as PM2.5) were taken. The locations were adjacent to major sources of air pollution, including waste burning at a dumpsite, motor vehicles, and cooking fires. PM2.5 concentrations are virtually always above 10 g $m^{-3}$ (the WHO yearly limit) and frequently exceed 25 g $m^{-3}$ at all sites (WHO 24-hour limit). These concentrations are higher than those found in European cities but lower than those found in Asian cities [10]. Gaseous pollutants (ozone $O_3$, nitrogen dioxide $NO_2$, sulphur dioxide $SO_2$) have no long-term observations in southern West African cities. DACCIWA conducted bi-monthly surface observations at the four air quality measuring sites from 2015 to 2017, as well as airborne observations in the summer of 2016. The concentrations of these contaminants did not surpass WHO guidelines. However, it is possible that $NO_2$ levels will rise on some days [11]. In Côte d'Ivoire, DACCIWA took direct measurements of particle and organic gas emissions from individual automobiles. They were much greater than the region's average. Old gasoline vehicles pollute the environment by a factor of a thousand. The performance of older diesel automobiles was just a factor of five worse. The emissions from new four-stroke engines are much lower than those from the new two-stroke engines [12].

For the first time, DACCIWA researched how the local population is affected in the cities of Abidjan and Cotonou. The number of hospital visits and PM2.5 concentrations was found to be significantly correlated at all three monitoring locations in Abidjan, especially during the rainy (summer) season. This shows that humidity may play an important role in the relationship between particulate matter and health, presumably by assisting in the inhalation of contaminants. The connections we detect between particulate matter and health outcomes vary by metropolitan region, implying that when addressing air quality implications on health, not only the concentration levels but also the source of PM2.5 should be considered. Long-term relative risk estimates for each municipality in Abidjan were derived using the number of medical visits as a benchmark for bad health outcomes. The link between long-term exposure to PM2.5 and respiratory, cardiac, and dermatologic health, as well as emergency room mortality, is described in this study. With PM2.5 concentrations decreased to the WHO recommended limit of 10 g $m^3$, the number of visits to the emergency department for respiratory or cardiac difficulties might be reduced by 3–4%, and up to 4% of emergency room mortalities could be averted [9]. Domestic fires pose a significant health danger because of the high concentration levels, although the risks from heavy traffic or rubbish burning were less severe. The findings may be obscuring the serious risk associated with long periods of time near a significant emission source because this study focused more broadly on residents of the neighbourhoods surrounding the DACCIWA measuring sites, rather than specifically on bus drivers, people working in food preparation, or at the landfill site. Direct contact assessments on various groups of people in the vicinity of these sites revealed that the health risk was highest for children in waste burning sites owing to heavy metals, while the risk was highest for women in the home burning site in the summer due to organic matter [9].

## 3. Conclusion

In general, air quality monitoring, public policy, and regulation in ECOWAS cities are improving slowly, while it still has a long way to go before reaching Europe, America, Asia, or even other African regions. Research must be encouraged to accurately measure and manage the origins and health impacts of urban air pollution in ECOWAS [3]. While this study has shed light on the major flaws in current inventories; long-term observational data is required to thoroughly evaluate emission inventories for the region and to allow for ongoing monitoring of emissions changes over time. The West African region currently lacks these observational data, and future studies should focus on developing a broader observational network in the region [4]. However, [13] believes that when it comes to investigating air pollution and its health impacts, Sub-Saharan Africa—and the current study has revealed that ECOWAS is acutely susceptible and is systematically left out of such studies, further lagging the region. Creating public awareness is particularly challenging due to a lack of air quality or health data to emphasise the seriousness of the situation. Investments in infrastructure, such as waste management and energy provision, are also important aspects of any holistic policy in these metropolitan contexts. Also, because facts alone do not form a policy, politicians and administrations must pay close attention and have a comprehensive vision to ensure successful urban air quality management, as updating laws is a critical first step. Some West African countries have environmental rights included in their constitutions, while others have air quality rules dating back to the 1980s and 1990s. Despite this solid start, many laws are already antiquated, with PM limits well exceeding WHO norms, for example. As a result, even in countries where standards are publicised, public health is not adequately protected [3]. To these limited regulations, issues of unstable governance and insufficient enforcement capacities must be added—even when legislation is established, maintaining compliance is typically difficult due to limited administrative resources. This is due in part to the fact that urban air pollution is a problem that impacts a wide range of industries. As a result, its administration necessitates a high level of coordination and cooperation amongst them. Although extensive evaluation studies are missing, having an organisation responsible for centralising air quality control, as in the examples of Senegal, Ghana, and the Ivory Coast, could lead to beneficial results in this regard [3].

In summary, it is not enough to simply provide these countries with air quality data; it is also critical to encourage long-term efforts that will boost data availability, good communication across government sectors and stakeholders, and ECOWAS-wide collaboration. This will increase government action, sufficient legislation, and public awareness, resulting in the area as a whole prioritising urban air quality and public health protection [3].

## Author details

Odubanjo D. Adedolapo
Department of Civil Engineering, Covenant University, Ota Ogun State, Nigeria

*Address all correspondence to: odubanjodolapo05@gmail.com

## References

[1] Fisher S, Bellinger DC, Cropper ML, Kumar P, Binagwaho A, Koudenoukpo JB, et al. Air pollution and development in Africa: Impacts on health, the economy, and human capital. The Lancet Planetary Health. 2021;**5**(10):e681-e688. DOI: 10.1016/S2542-5196(21)00201-1

[2] Abera A, Friberg J, Isaxon C, Jerrett M, Malmqvist E, Sjöström C, Taj T, Vargas AM. Air Quality in Africa: Public Health Implications. Annual Review of Public Health. 2020; 42L193-210. 10.1146/annurev-publhealth-100119-113802

[3] Alvarez CM, Hourcade R, Lefebvre B. A scoping review on air quality monitoring, policy and health in West African cities. International Journal of Environmental Research and Public Health. 2020;**17**(23):9151

[4] Morris ER. Simulations of Air Quality in West Africa: An Evaluation of the Emissions, Sources Seasonality and Impacts of Pollutants. Heslington, York, United Kingdom: University of York; 2019. Available from: http://etheses.whiterose.ac.uk/27056/

[5] Aliyu YA, Botai JO. Reviewing the local and global implications of air pollution trends in Zaria, Northern Nigeria. Urban Climate. 2018;**26**: 51-59

[6] Ngom B, Seye MR, Diallo M, Gueye B, Drame MS. A hybrid measurement kit for real-time air quality monitoring across Senegal Cities. In: Proceedings of the 2018 1st International Conference on Smart Cities and Communities (SCCIC), Ouagadougou, Burkina Faso, 24-26 July. California State University, Fullerton: SCCIC; 2018. pp. 1-6

[7] Toure N, Guèye ND, Diokhane AM, Jenkins G, Li M, Drame M, et al. Observed and modeled seasonal air quality and respiratory health in Senegal during 2015 and 2016. GeoHealth. 2020;**3**:423-442

[8] Knippertz P, Coe H, Chiu JC, Evans MJ, Fink AH, Kalthoff N, et al. The dacciwa project: Dynamics-aerosol-chemistry-cloud interactions in West Africa. Bulletin of the American Meteorological Society. 2015;**96**(9):1451-1460. DOI: 10.1175/BAMS-D-14-00108.1

[9] Evans MJ, Knippertz P, Aristide A, Allan RP. Policy-relevant findings of the DACCIWA Project. Rhode Island, USA: AMS; 2018. DOI: 10.5281/zenodo.1476843

[10] Djossou J, Léon J, Akpo AB, Liousse C, Yoboué V, Bedou M, et al. Mass concentration, optical depth and carbon composition of particulate matter in the major southern west African cities of Cotonou (Benin) and Abidjan (Côte d'Ivoire). Atmospheric Chemistry and Physics. 2018;**18**(9):6275-6291

[11] Bahino J, Yoboué V, Galy-Lacaux C, Adon M, Akpo A, Keita S, et al. A pilot study of gaseous pollutants measurement ($NO_2$, $SO_2$, $NH_3$, $HNO_3$ and $O_3$). Atmospheric Chemistry and Physics. 2018;**18**(7):5173-5198

[12] Keita S, Liousse C, Yoboué V, Dominutti P, Guinot B, Assamoi E, et al. Particle and VOC emission factor measurements for anthropogenic sources in West Africa. Atmospheric Chemistry and Physics. 2018;**18**(10):7691-7708

[13] Amegah AK. Proliferation of low-cost sensors. What prospects for air pollution epidemiologic research in sub-Saharan Africa. Environmental Pollution. 2018;**241**:1132-1137

Chapter 3

# Effects of a Saharan Dust Episode on Emergency Attendances for Respiratory Diseases in İstanbul

*Özkan Çapraz and Ali Deniz*

## Abstract

Saharan dust events have an important effect on the air quality of Turkey due to their significant contribution to particulate matter concentrations. These events likely impact public health in urban areas. However, there is no available information on the health effects of Saharan dust in Turkey. On Sunday 1 February 2015, İstanbul experienced an episode of extreme Sahara dust event, which increased the particulate matter concentrations greatly compared to the average values of the city. In this study, we examined the relationship between particulate matter ($PM_{10}$, $PM_{2.5}$) concentrations and emergency attendances for asthma, chronic obstructive pulmonary disease (COPD), and acute bronchitis on the episode day to better understand the association between an extreme dust event and emergency attendances for respiratory health in the city. Analyses showed that there was no significant effect of the Saharan dust event on emergency attendances for asthma and COPD in İstanbul compared to average emergency attendance numbers of the city. However, emergency attendances for acute bronchitis significantly increased on the episode day. This study revealed, extreme Saharan dust events can considerably increase the risk of visiting hospital for acute bronchitis in İstanbul during a severe dust episode.

**Keywords:** air pollution, Saharan dust episode, particulate matter, health effects, emergency attendances

## 1. Introduction

Large quantities of dust from arid lands and deserts can be transported thousands of kilometers by winds. Therefore, the impact of dust storms can be observed many hundreds of kilometers downwind from the emission source. Ambient concentrations of particulate matter during dust transportation events can be extremely high due to millions of tons of mineral dust and many people are likely to inhale a higher than usual concentration of particulate matter (PM) during these events. Studies suggest that dust storms can increase the risk of respiratory health effects, such as emergency department visits, hospital admissions, and mortality [1–5].

Sahara dust is the largest, naturally occurring source of particulate matter in The World. Due to the proximity to the Sahara Desert, Sahara dust transportation have an important effect on the air quality of Turkey and can lead to high PM levels that can exceed the limit values [6–9]. These events likely impact public health in urban areas in Turkey. However, there is no available information on the health effects of Saharan dust in Turkey so far.

On Sunday, 1 February 2015, İstanbul experienced a serious Sahara dust transport event caused by strong southwesterly winds. During this episode, the sky turned dusty orange, ferry traffic was interrupted, and over 100 flights were canceled by Turkish Airlines. Daily average $PM_{10}$ concentration was recorded at 325,1 $\mu g/m^3$, which is an extremely high level of ambient $PM_{10}$ not normally encountered in the city. This episode provided an opportunity to assess the health effects during an extreme dust event that significantly affected daily city life. We conducted an analysis to determine the effects of this episode on the emergency department visits for asthma, COPD, and acute bronchitis in İstanbul by using air quality and health data. Synoptic analysis and NASA satellite maps are also used to show the dust transportation from the Sahara Desert as a dust source region.

## 2. Methodology

### 2.1 Study area

Located between Black Sea and Sea of Marmara, İstanbul is the largest urban area of Turkey with a population of more than 16 million. The city forms one of the largest urban agglomerations in Europe. İstanbul is separated into Asian and European parts by Bosphorus strait which is approximately 30 km in length.

### 2.2 Air quality data

Hourly air pollution data ($PM_{10}$ and $PM_{2.5}$) were used from the database of Republic of Turkey Ministry of Environment and Urbanization, the government agency in charge of collection of air pollution data in Turkey. The hourly concentrations for each pollutant were obtained from the 9 fixed-site air quality monitoring stations of Marmara Clean Air Center Monitoring Network (Kağıthane, Esenyurt, Silivri, Ümraniye, Mecidiyeköy, Şirinevler, Üsküdar and Kandilli). To evaluate the weather conditions on the episode days, hourly meteorological data (temperature and relative humidity) were obtained from the Air Quality Monitoring Stations where meteorological measurements are also made (**Table 1**). Daily means of the concentrations calculated from the hourly data of the pollutants and weather variables were used to represent the daily reading for İstanbul.

### 2.3 Satellite data

In this study, Hybrid Single-Particle Lagrange Integrated Trajectory Model developed by Air Resources Laboratory of National Oceanic and Atmospheric Administration (NOAA) was used to detect the trajectories showing dust transportation during the episode. Region coordinates are marked on the map in HYSPLIT model and past trajectories showing dust transportation on the maps are obtained. In order to visually examine the high particulate matter concentrations during the episode day, Moderate Resolution Imaging Spectroradiometer (MODIS) on NASA's Aqua satellite images were also used in the study.

| | Mean + SD | Min | P(25) | P(50) | P(75) | Max | Episode Day |
|---|---|---|---|---|---|---|---|
| Number of emergency attendances | | | | | | | |
| Acute bronchitis (n = 154,182) | 369 ± 200 | 51 | 228 | 337 | 471 | 1259 | 829 |
| Asthma (n = 65,592) | 157 ± 59 | 52 | 122 | 155 | 186 | 779 | 153 |
| COPD (n = 25,240) | 60 ± 24 | 15 | 42 | 59 | 74 | 217 | 33 |
| Air pollution concentrations (μg/m3) | | | | | | | |
| $PM_{10}$ | 52.7 ± 24.7 | 17.0 | 35.3 | 45.6 | 60.5 | 325.1 | 325.1 |
| $PM_{2.5}$ | 25.5 ± 14.6 | 5.6 | 15.4 | 20.9 | 31.4 | 105.0 | 70.6 |
| Weather | | | | | | | |
| Temperature (°C) | 15.2 ± 7.1 | −3.8 | 9.5 | 15.1 | 22.0 | 30.0 | 17,4 |
| Humidity (%) | 79.5 ± 10.7 | 45.9 | 72.7 | 80.2 | 86.9 | 100 | 50,6 |

**Table 1.**
*Summary statistics of number of emergency attendances, air pollutant concentrations, and weather conditions in İstanbul, Turkey (2014–2017).*

### 2.4 Health data

Data on the number of daily respiratory hospital admissions of the 21 public hospitals in İstanbul from January 1, 2014, to December 31, 2017 (1445 days) were obtained from the database of Republic of Turkey Ministry of Health. The daily respiratory hospital admission counts in our study were selected for seven different respiratory outcomes according to the International Classification of Disease, Tenth Revision (ICD-10) by the World Health Organization: Acute bronchitis (J20), COPD (J44), and asthma (J45-J46).

In Turkey, admissions for state hospitals (except emergency services) on weekdays are made by appointment taken via the internet. Weekend services of state hospitals are limited to emergencies services only. In this study, due to the Saharan dust episode that occurred on Sunday, all the hospital admissions made on the day of the episode are attendances to emergency services.

## 3. Results

On 01 February 2015, a cyclone centered on Italy caused a southerly wind event on the eastern Mediterranean (**Figure 1**). Sahara Desert sands were lifted by strong winds, passed over Aegean Sea, and reached İstanbul. Dust load transport and its path can be seen well on NASA Earth Data imaginary as a continuous dust flow from North Africa to the Black Sea (**Figure 2**). HYSPLIT 48 hr. air-mass back trajectories at different elevations (**Figure 3**) indicate the movement of air masses from Saharan Desert in North Africa passing through Mediterranean to İstanbul.

The stations of air quality monitoring network in İstanbul measured high concentrations of $PM_{10}$ (325.1 μg/m$^3$) and $PM_{2.5}$ (70.6 μg/m$^3$) on February 1, 2015, which were good matched with the NASA satellite image and HYSPLIT model trajectories. Particulate matter size distribution was dominant at a 3–7.2 μm size fraction on the episode day [13]. On February 2, dust transportation was ceased, and precipitation started. Consequently,

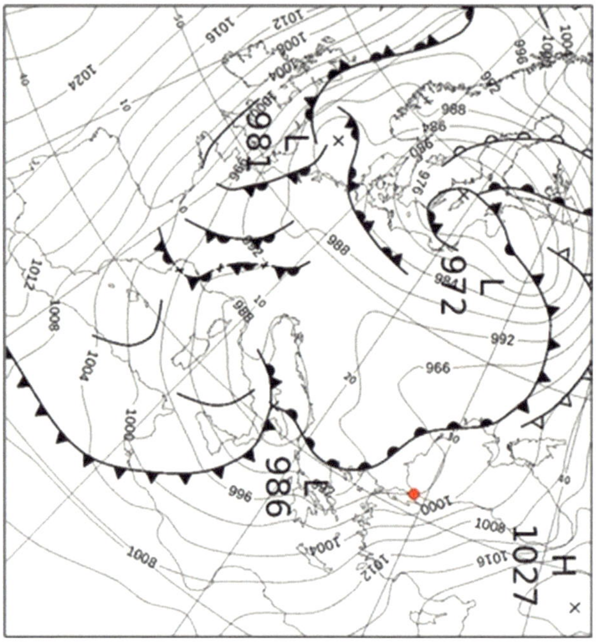

**Figure 1.**
*Mean sea-level pressure (hPa) map of Europe showing weather fronts on 01 February 2015 [10]. İstanbul is marked with a red dot.*

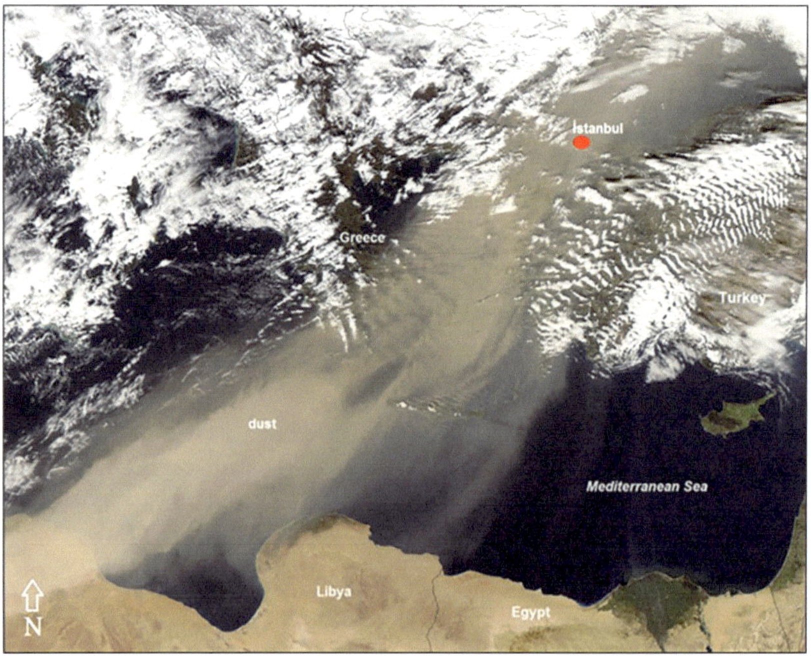


**Figure 2.**
*Saharan dust transport on 01 February 2015 [11].*

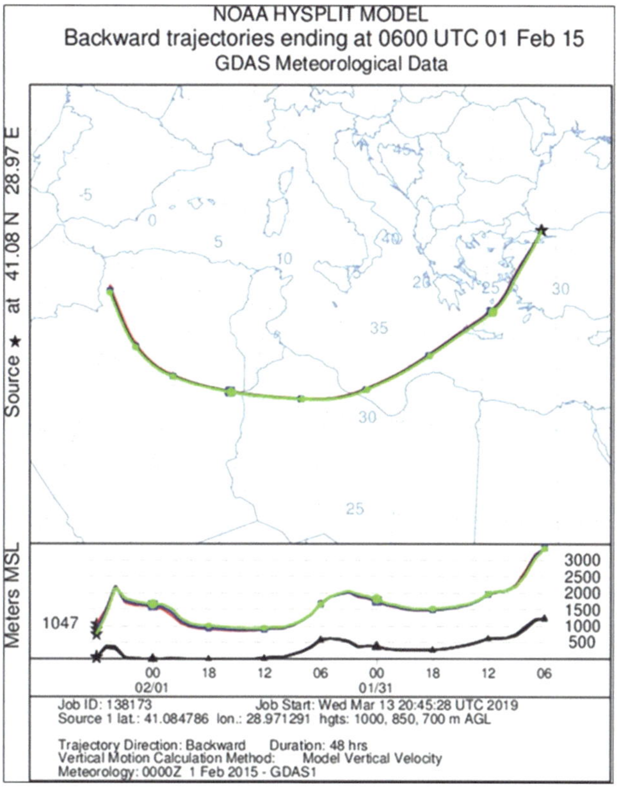


**Figure 3.**
*HYSPLIT backward air mass trajectories originating from North Africa and arriving in Istanbul on 01 February 2015 [12].*

particulate matter concentrations decreased with wet and dry deposition. **Figure 4** shows the significant contribution of Saharan dust to ambient concentrations of particulate matter during the episode. Descriptive statistics for daily mean pollutant levels, temperature, and humidity over a 4-year period between 2014 and 2017 are presented in **Table 1**, showing comparisons of concentrations on the dust storm day (01/02/15) with the average numbers between 2014 and 2017. The $PM_{10}$ concentration recorded on the episode day (325.1 $\mu g/m^3$), which greatly exceeded the European union daily standard of 50 $\mu g/m^3$, is the highest concentration of daily average $PM_{10}$ observed between 2014 and 2017 in İstanbul. Yearly the average $PM_{10}$ concentration was 52.6 $\mu g/m^3$ between 2014 and 2017 in the city, respectively. Considering the yearly average, it could be inferred that ambient $PM_{10}$ concentration increased more than 6 times during the dust episode.

The $PM_{2.5}$ concentration recorded on the episode day was 70.6 $\mu g/m^3$. Yearly average $PM_{2.5}$ concentrations between 2014 and 2017 were 25.5 $\mu g/m^3$ (**Table 1**). Although a high $PM_{2.5}$ value is measured on the episode day, $PM_{2.5}$ concentration did not rise considerably compared to the very high $PM_{10}$ value (**Figure 4**). Low atmospheric pressure, warm air, and strong winds were observed during the episode. The daily average air temperature on the day of the event was 17.4°C and the daily average wind speed (6.5 m/s) was also very high compared to the average temperature (6.1°C) and wind speed values (2.6 m/s) of the city (**Table 1**).

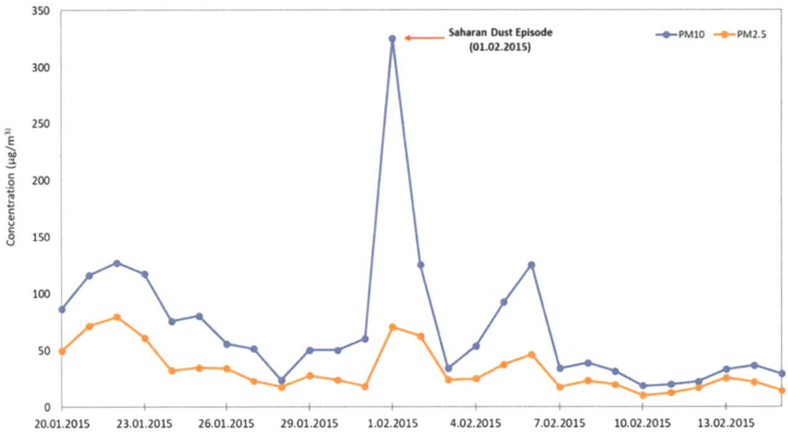


**Figure 4.**
*Daily mean values of $PM_{10}$ and $PM_{2.5}$ measured in İstanbul between 20/01/2015 and 15/02/2015 [9].*

Descriptive statistics for daily mean emergency department visits for three different respiratory diseases are presented in **Table 1**, showing comparisons of the ED visits on the dust episode day (01/02/15) with the average ED admissions between 2014 and 2017. Because the Saharan dust episode happened on Sunday, all hospital applications on the episode day are made for emergency departments. According to our results, the dust storm episode had no significant effect on the emergency department visits for asthma and COPD; ED visits for asthma (n = 153) and COPD (n = 33) on the episode day were below the average values of ED visits (avg. = 157 and avg. = 60) made between 2014 and 2017 (**Table 1**). However, the dust storm episode had a significant effect on the emergency department visits for acute bronchitis (**Figure 5**). Elevated Saharan dust levels exacerbated acute bronchitis in patients, as assessed by the increasing number of ED admissions (n = 829) on 01/02/15. According to results, more than twice of the average ED visits (avg. = 369) have been made on the episode day (**Table 1**).

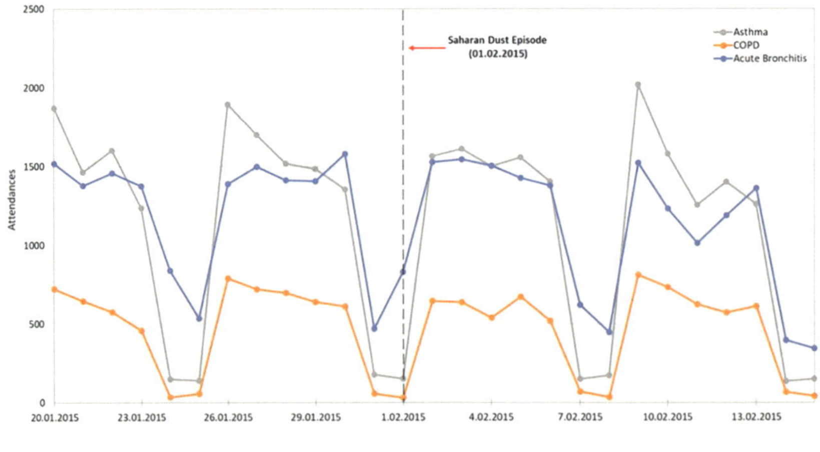


**Figure 5.**
*Hospital applications for asthma, COPD, and acute bronchitis in İstanbul between 20/01/2015 and 15/02/2015.*

## 4. Discussion and conclusion

In this study, the relationship of a severe Saharan dust episode with the emergency department visits for asthma, chronic obstructive pulmonary disease (COPD), and acute bronchitis in İstanbul was examined. Investigation of the impact of a Saharan dust episode on respiratory diseases has not been done previously in Turkey, yet. Our results showed that the dust episode did not significantly affect ED visits for asthma and COPD as there were no substantial changes on the emergency visits that could be linked to an extreme increase in dust concentration. However, the dust storm was significantly associated with increased ED visits for acute bronchitis as there was a large increase in ED visits on the episode day compared to average levels of the city.

Many studies stated that dust storms can increase the risk of emergency department visits, respiratory hospitalizations, or admissions for respiratory diseases. A study conducted in Guadeloupe Island over 1 year found that $PM_{10}$ and $PM_{2.5–10}$ pollutants contained in the Saharan dust increased the risk of visiting the health emergency department for children with asthma [14]. In Rome, Alessandrini et al. [1] investigated the residents hospitalized during Saharan dust days between 2001 and 2004. They found that the effect of $PM_{2.5–10}$ on respiratory hospitalizations was higher during dust days compared with dust-free days. No effect of $PM_{2.5}$ was detected. Saharan dust also increased the effect of $PM_{10}$ on cerebrovascular diseases. Extremely high levels of particulate matter were recorded, with daily average levels of coarse matter (<10 μm) peaking over 11,000 μg/m$^3$ and fine (<2.5 μm) over 1600 μg/m$^3$ d. in Sydney, Australia. The dust storm period was associated with large increase in asthma emergency department visits, and respiratory emergency department visits. There was no significant increase in cardiovascular emergency department visits or hospital admissions [14]. According to a study in Taiwan, Asian dust storms increase cardiopulmonary emergency visits in Taipei when ambient $PM_{10}$ concentrations are above 90 μg/m$^3$. Compared to their pre-dust periods, emergency visits for ischemic heart diseases, cerebrovascular diseases, and COPD during high dust events are increased by 0.7 cases (35%), 0.7 cases (20%), and 0.9 cases (20%) per event, respectively [15].

However, several studies did not find any link between dust episodes and negative health outcomes. Prospero et al., found no substantial changes in pediatric asthma attendances that could be linked to short-term surges in dust concentrations from African dust carried in the Atlantic Trade Winds in Barbados reporting on a 2-year study [16]. In a study undertaken in Saudi Arabia, where major sandstorms are frequent, there was no correlation between the average daily $PM_{10}$ and $PM_{2.5}$ levels and the average number of children presenting with acute asthma per day, their daily asthma score, or admission rate [17]. The study of Menendez et al. [18] conducted in Gran Canaria (Spain) found no statistically significant relations between emergency patients with respiratory diseases and elevated Saharan dust levels.

Considering the health effects of particulate matter, it is suggested to investigate the effects of dust transport episodes on human health in future studies.

## Conflict of interest

"The authors declare no conflict of interest."

## Author details

Özkan Çapraz* and Ali Deniz
İstanbul Technical University, İstanbul, Turkey

*Address all correspondence to: caprazozkan@gmail.com

## References

[1] Alessandrini ER, Stafoggia M, Faustini A, Gobbi GP, Forastiere F. Saharan dust and the association between particulate matter and daily hospitalisations in Rome, Italy. Occupational and Environmental Medicine. 2013;**70**(6):432-434

[2] Johnston F, Hanigan I, Henderson S, Morgan G, Bowman D. Extreme air pollution events from bushfires and dust storms and their association with mortality in Sydney, Australia 1994-2007. Environmental Research. 2011;**111**(6):811-816

[3] Hsieh NH, Liao CM. Assessing exposure risk for dust storm events-associated lung function decrement in asthmatics and implications for control. Atmospheric Environment. 2013;**68**:256-264

[4] Goudie AS. Desert dust and human health disorders. Environment International. 2014;**63**:101-113

[5] Zhang X, Zhao L, Tong DQ, Wu G, Dan M, Teng B. A systematic review of global desert dust and associated human health effects. Atmosphere. 2016;7(12):158

[6] Baltacı H. Spatial and temporal variation of the extreme Saharan dust event over Turkey in march 2016. Atmosphere. 2017;**8**(2):41

[7] Kabatas B, Pierce RB, Unal A, Rogal M. April 2008 Saharan dust event: Its contribution to PM10 concentrations over the Anatolian peninsula and relation with synoptic conditions. Science of the Total Environment. 2018;**633**:317-328

[8] Özdemir T. Investigations of a southerly non-convective high wind event in Turkey and effects on PM10 values: A case study on April 18, 2012. Pure and Applied Geophysics. 2019;**176**:4599-4622

[9] Çapraz Ö, Deniz A. Particulate matter (PM10 and PM2.5) concentrations during a Saharan dust episode in Istanbul. Air Quality, Atmosphere and Health. 2021;**14**:109-116

[10] Wetter3. UKMET analysis charts. 2021. Available from: http://www1.wetter3.de/archiv_ukmet_en.html

[11] NASA. EOSDIS Worldview. 2021. Available from: https://worldview.earthdata.nasa.gov/

[12] NOAA. HYSPLIT Model. 2021. Available from: https://www.ready.noaa.gov/HYSPLIT.php

[13] Kuzu SL. Compositional variation of PCBs, PAHs, and OCPs at gas phase and size segregated particle phase during dust incursion from the Saharan Desert in the northwestern Anatolian peninsula. Advances in Meteorology. 2016;**2016**:1-12

[14] Merrifield A, Schindeler S, Jalaludin J, Smith W. Health effects of the September 2009 dust storm in Sydney, Australia: Did emergency department visits and hospital admissions increase? Environmental Health. 2013;**12**:32

[15] Chan CC, Chuang KJ, Chen WJ, Chang WT, Lee CT, Peng CM. Increasing cardiopulmonary emergency visits by long-range transported Asian dust storms in Taiwan. Environmental Research. 2008;**106**(3):393-400

[16] Prospero JM, Blades E, Naidu R, Mathison G, Thani H, Lavoie MC. Relationship between African dust

carried in the Atlantic trade winds and surges in pediatric asthma attendances in the Caribbean. International Journal of Biometeorology. 2008;**52**(8):823-832

[17] Alangari AA, Riaz M, Mahjoub MO, Malhis N, Al-Tamimi S, Al-Modaihsh A. The effect of sandstorms on acute asthma in Riyadh, Saudi Arabia. Annals of Thoracic Medicine. 2015;**10**(1):29-33

[18] Menendez I, Derbyshire E, Carrillo T, Caballero E, Engelbrecht PJ, Romero LE, et al. Saharan dust and the impact on adult and elderly allergic patients: The effect of threshold values in the northern sector of gran Canaria, Spain. International Journal of Environmental Health Research. 2017;**27**(2):144-160

Section 2

# Air Quality and COVID-19 Pandemic

Chapter 4

# COVID-19, Air Pollution and One Health at the Climate Change Turning Point

*Riccardo Pansini and Lei Shi*


## Abstract

COVID-19 escalated into a pandemic posing humanitarians and scientific challenges. We explored the geographical feature of the first wave infection and correlated it with annual satellite and ground indexes of air quality in eight countries: China, U.S.A, Italy, Iran, France, Spain, Germany, and U.K. Controlling for population size, we found more viral infections in those areas which were afflicted by high PM 2.5 and nitrogen dioxide values. Higher mortality was also correlated with relatively poor air quality. This phenomenon also occurs in China when removing, the city of Wuhan and its province from the dataset. For long recognised to be a high-risk factor for several respiratory-related diseases and conditions, air pollution seems to be a risk factor for COVID-19 too. This finding suggests the detrimental impact climate change will have on the trajectory of future respiratory epidemics. Previous Asian epidemics and the Ebola have brought forward evidence of the natural causes of zoonoses which have become more threatening due to land-use change, ensued lack of a buffer zone between the cities and the forests, and our closer proximity to wild pathogens. Together with air pollution, these elements illustrate the need to stick to the UN targets limiting biodiversity loss and climate change.


**Keywords:** air pollution, COVID-19, risk factor, zoonosis, climate change

## 1. Introduction

In early 2020, the world was hit by a pandemic of epochal dimensions as never seen before since a century earlier when the 1918 Spanish influenza hit worldwide. Too much of the public concern, such a global, new disease brought about several complot theories. Typical to difficult to explain phenomena, humans' resort to an intelligent hand as the causing factor behind them. In fact, during these last 20 years, the SARS outbreak from Southern China and Hong Kong, the Zika one from Central and South America, and the Ebola one from West Africa have shown how very large epidemics can originate in every continent with similar patterns.

Like the other SARS coronaviruses, scientists have provided the best evidence that SARS-CoV-2 transferred hosts from the originating reservoir of bats to humans [1] without any laboratory modification. Despite some warnings that SARS-CoV-2 might have occurred before its allegedly first outbreak in Wuhan, the most likely place where the **spillover** took place remains that wet market in Wuhan. To environmental scientists, though, this is an unimportant, episodic event that will always tell just one piece of a bigger story composed by the many accounts of cross-species transfers. In the last few years, these accounts of new viral zoonotic diseases have repeated and what they brought have been harmful consequences, such as the other of the three most relevant coronaviruses, the MERS from the middle-east in which dromedaries were involved [2].

Regardless of the specific epidemics, we look at, we know that in these last epidemics both the causes of the spill overs and later the fast spreads across different habitats and cities were all anthropogenic. Land-use change and habitat loss together with the **biodiversity loss** causing the sixth mass extinctions are the climate change reasons behind epidemics [3, 4]. As the WWF neatly summarises (**Figure 1**) [5], as long as the forest ecosystems are healthy and there exists a proper countryside buffer (without intensive farming) to the cities, spill overs to humans are extremely unlikely [6].

**Intensive livestock farming** is another key element contributing to zoonotic diseases [7] that coupled with human overpopulation, played a critical role [8]. Evidence flourished showing that a higher frequency of contagions occurred in German [9] and French abattoirs [10]. A multidisciplinary ecological approach to study the coronavirus and other zoonotic infections is necessary as well as combining medical factors causing co-morbidity before and after the disease in patients more at risk.

COVID-19 appeared in a Chinese area affected by some of the highest air pollutions in the world, and it showed a relatively high virulence there. The areas in the world that were particularly hit by the pandemic were those with high commercial trade [11] and those hit by air pollution. These are **risk factors** that, together with

**Figure 1.**
*WWF pictorial.*

medical issues including smoking, get associated with higher morbidity and mortality of COVID-19 [12]. In particular, smoking affects hyperactivates the ACE20 receptors in our alveolar respiratory system which are also attacked by the coronavirus [13]. Chronic exposure to air pollution causes a chronic lung and bronchial infection similar in features as that found in smokers because of the ACE20 receptors [14, 15].

## 2. Air pollution and COVID-19: the case of 8 studied countries

In late 2020 winter, alarmed by the spreading of the virus from China to Iran and Italy, we were struck by how it was hitting the most polluted region of the European continent, the Italian Po′ river plane. Of course, air pollution was just one of the reasons of the virus spread, mostly the travelling of **passengers** from China to Mecca, which brought it to Iran, and later of Chinese merchants that went to trading exhibitions in Italy were the first causes of the spread of the disease out of China. Yet, in China itself, SARS-CoV-2 was first recorded in one of the most polluted areas in the world. Conversely and paradoxically instead, the ecologically lush Chinese Yunnan Province, the one from where most of the coronavirus bat hosts have ever been sampled [16], has been one of the areas in the world's least affected by the pandemic. A zoonotic disease was seemingly hitting again in areas of the world where unsustainable human practices were present likely because of wild animals trading in a wet market.

We, therefore, started working at the theory that chronic exposure to air pollutants was a cofactor for higher cases and deaths from COVID-19 [17]. This was the safer **long-term air pollution hypothesis**, while confronting some experts in the field we understood others had the same intuition and, amongst those, bold physicists and environmental doctors were working on the more alarming short-term hypothesis that the virus is also carried by the polluting particulate matter [18]. Colleagues from America showed the same day of our preprint that other individuals and environmental risk co-factors play a minor role than pollution [19]. We were glad because the same idea was picked up by scientists with more experience than ourselves, and we enlarged the analysis to include the other countries of Spain, the U.K., the U.S.A., France, and Germany [20]. Later, dozens of other studies showed a similar trend occurring for many pollutants in many regions of the world.

What was important for our analyses to work was to find territories small enough to split the countries from which we could compare **COVID-19 data** and 2019 averaged pollution data. We repeated the analysis throughout the months of 2020 and it was frustrating not to be able to obtain, for example, deaths at a small provincial level from Italy; the country was not disclosing this important piece of information to the scientific community! We were glad to be able to use very important and early infection data published from the Iranian state in Farsi; those data stopped being published just after a few months and the country at a larger regional level could not be followed in its progression.

Relying on the expertise of the satellite geographer and co-author Davide Fornacca, the data on air pollution posed less trouble. Not totally trusting the accuracy of ground measures collected from the various devices around the world, we relied on data sampled from a European open-access **satellite** that estimates ground pollution around the globe at a fine scale.

The comparison between cases and deaths from COVID-19 and 2019 air pollution had to be controlled for the most important confounding factor within those comparably small territories in the 8 countries: **population density**. The higher the densities of the new hosts, the higher the contagions were going to be regardless of the type of parasite and ecological conditions around; it is a simple transmission paradigm we were not interested to look at.

A plethora of other confounding **risk cofactors** could have been added, and many other colleagues did so at a smaller scale than eight countries. From our side, we were aiming at a simple and global illustrative variable that represents a degraded environment and is involved in respiratory disease; a variable that could have played an even more important role once a disease becomes endemic and infections from distant places play less of a role.

The **maps** (**Figure 2** reported from our main publication [21]) show how the most polluted regions overlap with the areas where COVID-19 infects or causes most deaths. The trend is most evident in Italy, where we find a marked gradient between a highly polluted and the majority of cleaner areas. Germany and Spain provided no clear evidence of this correlation because of the character of the air pollution there. In Germany, pollution is evenly spread and the lack of a gradient does not allow for a Kendall correlation to output significant results. In Spain the pollution is negligible and the statistics also do not work out at a country level. Yet, later accounts could find some evidence also for these countries (as reviewed by [22] and later reviewed by more recent papers referenced at the end of this paper).

The **pollutants** we analysed were several, and amongst the most influential for finding a trend there are PM 2.5 (of the previous **Figure 2**) and PM 10 and, reported in the following **Figure 3** (reported from the other publication [23]) for China, $NO_2$, CO, and HCHO).

The China 'case' drew much of our attention right because it was where the epidemic started and the country where it was contained most effectively. The pollution hypothesis should not have stood much because the onset of the virus should

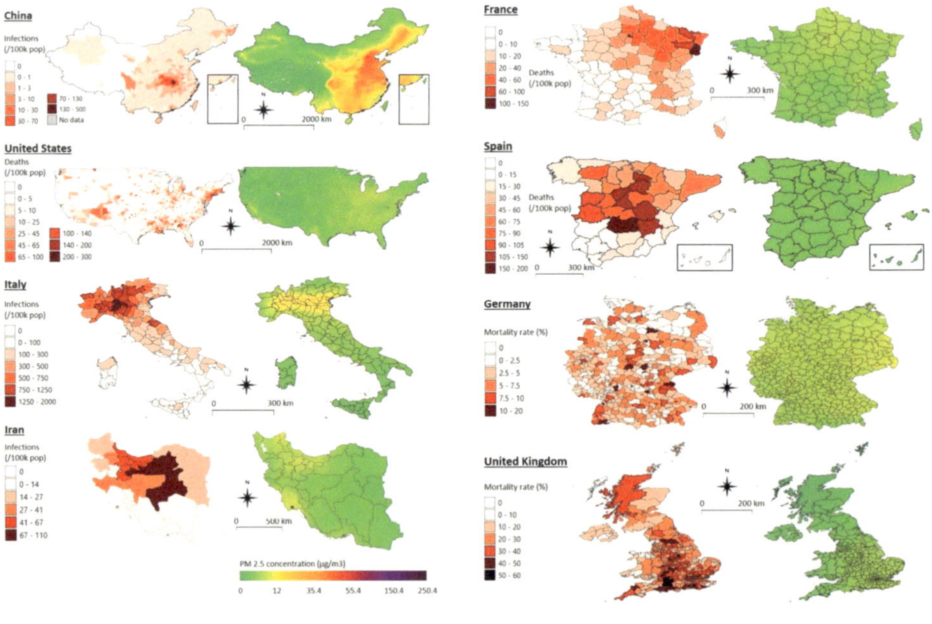


**Figure 2.**
*COVID/Pollution correlation maps of 8 countries.*

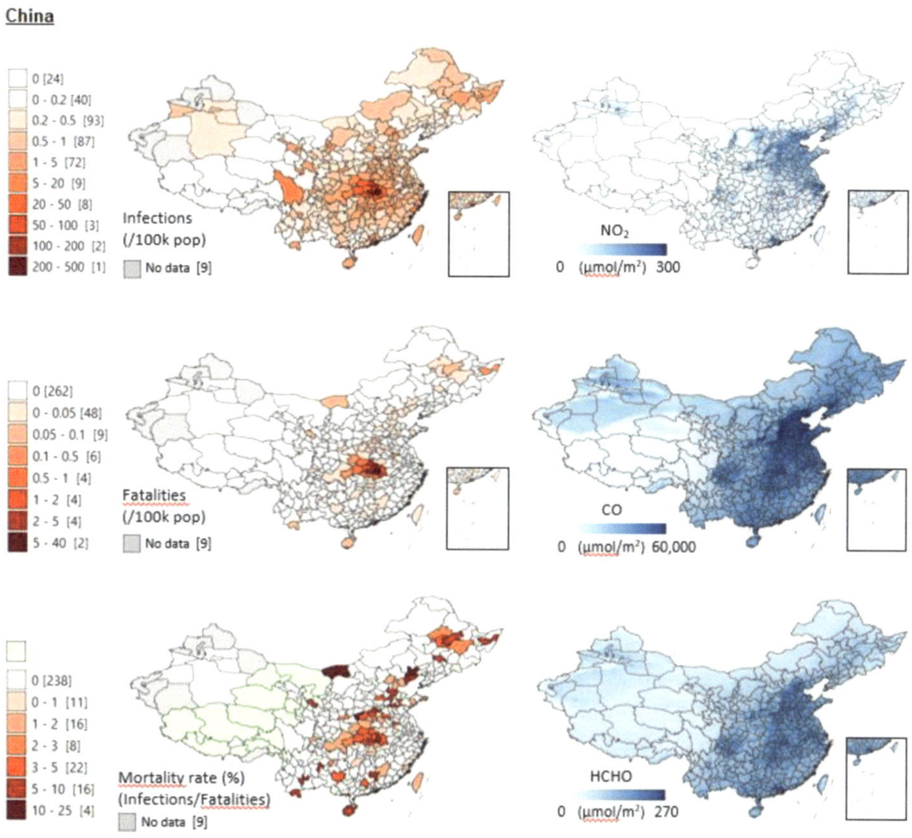


**Figure 3.**
*COVID/Pollution correlation maps for three pollutants in China.*

have been attributed to the location of the spill over. If it had happened somewhere with a high population density, that is in a megalopolis like Wuhan, pollution should have played a role possibly only later on from the initial abrupt explosion. Instead, the trend was there, even after removing Wuhan and its Hubei Province from the analysis [23].

Was the publication of these results easy in a high-impact journal? The earliest authors reflected on some limitations in a journal with high-impact factors [24]. We were unable to have such an impact for supposedly two or three reasons. Firstly, a correlational study is still an observational one missing an artificial intervention as such to be able with a control, non-modified group of patients, to compare results. Clear causation is, therefore, lacking. It is obvious that we cannot spread the virus ourselves in polluted and non-polluted places, as well as we cannot start polluting clean places for months to check whether, there, there will be more COVID cases and deaths. To overcome this ethical impossibility to turn the observation into an experiment, with years' time longitudinal screenings performed on patients from retrospective cohorts will help cast doubts out [25]. A first example has been attempted in Spain [26] with results which agree with out hypothesis [27]. The other **limitation** may be given by the truly interdisciplinary nature of the study which finds it hard to gather environmental medical doctors in general science journals who can also assess the geographical methods used here. The third, more important reason is given by the apparent nature of the problem; although air pollution like smoking is a problem causing millions of deaths every year, it is not big news... Nevertheless, the editors of journals and journalists with a large impact on the general public missed the chance to communicate the message at the right time of the epidemics, that is during the earliest months, when the scariest new deaths during the lockdowns could have allowed us to ponder on the true nature of these recent new diseases, including their ecological characteristics linked to climate change [28].

Several reviews collated our studies together with well over a hundred during almost 2 years. A few recent ones are listed in refs. [29–32].

## 3. Conclusions

The fossil fuel economy causing air pollution carried on unabated once we resumed activities after the lockdowns. At an individual level, we are fast at not wanting to tackle the emission problem because the institutions have not endorsed reforms effectively [33]. However, the basics of game theory tell us that we need to have a proper behavioural strategy at an individual level before expecting to see groups of individuals or institutions act likewise. As well as solving the Public Good Game of the pandemic [34] by paying a small individual cost and getting vaccinated for the benefit of the collectively, we must also more strongly clean our environment and strongly limit our impact for the good of the collectively and ourselves. The fact that nature is going to profit and that we need to pay a cost seems to cause us an extra layer of difficulty in the swift employment of these green strategies, when in fact, we do not understand that we belong to that layer too.

What saves the planet will save us too.

## Acknowledgements

Davide Fornacca contributed to accomplishing our preprints and the two main referenced papers published in the journals Atmosphere and Frontiers in Public Health [21, 23]. This chapter as well as these two previous works received no funding for their completion.

## Conflict of interest

The authors declare no conflict of interest.

*DOI: http://dx.doi.org/10.5772/intechopen.102943*

## Author details

Riccardo Pansini[1,2,3]* and Lei Shi[1,4]

1 Behavioral and Experimental Economics Research Center, Statistic and Mathematics College, Yunnan University of Finance and Economics, Kunming, China

2 Institute of Eastern-Himalaya Biodiversity Research, Dali University, Yunnan Province, China

3 Department of Economics and Finance, Global Research Unit, City University of Hong Kong, Hong Kong

4 Interdisciplinary Research Institute of Data Science, Shanghai Lixin University of Accounting and Finance, Shanghai, China

*Address all correspondence to: r.pansini@gmail.com

## References

[1] Andersen KG, Rambaut A, Lipkin WI, Holmes EC, Garry RF. The proximal origin of SARS-CoV-2. Nature Medicine. 2020;**26**(4):450-452. DOI: 10.1038/s41591-020-0820-9

[2] Killerby ME, Biggs HM, Midgley CM, Gerber SI, Watson JT. Middle East respiratory syndrome coronavirus transmission. Emerging Infectious Diseases. 2020;**26**:191

[3] Johnson CK, Hitchens PL, Pandit PS, Rushmore J, Evans TS, Young CCW, et al. Global shifts in mammalian population trends reveal key predictors of virus spillover risk. PRSB. 2020;**287**:20192736. DOI: 10.1098/rspb.2019.2736

[4] Gibb R, Redding DW, Chin KQ, Donnelly CA, Blackburn TM, Newbold T, et al. Zoonotic host diversity increases in human-dominated ecosystems. Nature. 2020;**584**(7821):398-402. DOI: 10.1038/s41586-020-2562-8

[5] Pratesi I, Galaverni M, Antonelli M. Pandemie, l'effetto boomerang della distruzione degli ecosistemi. Rome: WWF Italia Onlus; 2020. Availabe from: https://www.wwf.it/pandanews/ambiente/le-foreste-sono-il-nostro-antivirus-ma-le-stiamo-distruggendo/ [Accessed: 01 September 2021]

[6] Morand S, Owers K, Bordes F. Biodiversity and emerging zoonoses. In: Confronting Emerging Zoonoses. Berlin: Springer; 2014. pp. 27-41. DOI: 10.1007/978-4-431-55120-1_3

[7] Taylor CA, Boulos C, Almond D. Livestock plants and COVID-19 transmission. Proceedings of the National Academy of Sciences. 2020;**117**:31706-31715

[8] Bedford J, Farrar J, Ihekweazu C, Kang G, Koopmans M, Nkengasong J. A new twenty-first century science for effective epidemic response. Nature. 2019;**575**:130-136. DOI: 10.1038/s41586-019-1717-y

[9] Oltermann P. Abattoir Air Cooling Systems Could Pose Covid-19 Risks, Expert Warns. Manchester, UK: The Guardian; 2020. Availabe from: https://www.theguardian.com/world/2020/jun/25/abattoir-air-cooling-systems-could-pose-covid-19-risks-expert-warns

[10] Coronavirus: Plus De Cent Cas De Covid-19 Détectés Dans Deux Abattoirs Français. Le Monde, French; 2020. Availabe from: https://www.lemonde.fr/planete/article/2020/05/17/plus-de-cent-cas-de-covid-19-detectes-dans-deux-abattoirs-francais_6039960_3244.html

[11] Bontempi E. A global assessment of COVID-19 diffusion based on a single indicator: Some considerations about air pollution and COVID-19 spread. Environmental Research. 2022;**204**:112098. DOI: 10.1016/j.envres.2021.112098

[12] Umnuaypornlert A, Kanchanasurakit S, Lucero-Prisno DE, Saokaew S. Smoking and risk of negative outcomes among COVID-19 patients: A systematic review and meta-analysis. Tobacco Induced Diseases. 2021;**19**:01-13. DOI: 10.18332/tid/132411

[13] Milne S, Yang CX, Timens W, Bossé Y, Sin DD. SARS-CoV-2 receptor ACE2 gene expression and RAAS inhibitors. The Lancet Respiratory Medicine. 2020;**8**:e50-e51. DOI: 10.1016/S2213-2600(20)30224-1

[14] Paital B, Agrawal PK. Air pollution by $NO_2$ and PM 2.5 explains COVID-19 infection severity by overexpression

of angiotensin-converting enzyme 2 in respiratory cells: A review. Environmental Chemistry Letters. 2020;**19**(1):1-18

[15] Frontera A, Cianfanelli L, Vlachos K, Landoni G, Cremona G. Severe air pollution links to higher mortality in COVID-19 patients: The "double-hit" hypothesis. Journal of Infection. 2020;**81**(2):255-259. DOI: 10.1016/j.jinf.2020.05.031

[16] Han Y, Du J, Su H, Zhang J, Zhu G, Zhang S, et al. Identification of diverse bat alphacoronaviruses and betacoronaviruses in China provides new insights into the evolution and origin of coronavirus-related diseases. Frontiers in Microbiology. 2019;**10**:1900

[17] Pansini R, Fornacca D. Initial Evidence of higher morbidity and mortality due to SARS-CoV-2 in regions with lower air quality. medRxiv. 2020. DOI: 10.1101/2020.04.04.20053595

[18] Setti L, Passarini F, De Gennaro G, Barbieri P, Pallavicini A, Ruscio M, et al. Searching for SARS-COV-2 on particulate matter: A possible early indicator of COVID-19 epidemic recurrence. International Journal of Environmental Research and Public Health. 2020;**17**:2986. DOI: 10.3390/ijerph17092986

[19] Wu X, Nethery RC, Sabath BM, Braun D, Dominici F. Exposure to air pollution and COVID-19 mortality in the United States. medRxiv. 2020. DOI: 10.1101/2020.04.05.20054502

[20] Pansini R, Fornacca D. Higher virulence of COVID-19 in the air-polluted regions of eight severely affected countries. medRxiv. 2020. DOI: 10.1101/2020.04.30.20086496

[21] Pansini R, Fornacca D. Early spread of COVID-19 in the air-polluted regions of eight severely affected countries. Atmosphere. 2021;**12**:795. DOI: 10.3390/atmos12060795

[22] Comunian S, Dongo D, Milani C, Palestini P. Air pollution and Covid-19: The role of particulate matter in the spread and increase of Covid-19's morbidity and mortality. International Journal of Environmental Research and Public Health. 2020;**17**:4487. DOI: 10.3390/ijerph17124487

[23] Pansini R, Fornacca D. COVID-19 higher mortality in Chinese regions with chronic exposure to lower air quality. Frontiers in Public Health. 2021;**8**:999. DOI: 10.3389/fpubh.2020.597753

[24] Wu X, Nethery RC, Sabath MB, Braun D, Dominici F. Air pollution and COVID-19 mortality in the United States: Strengths and limitations of an ecological regression analysis. Science Advances. 2020;**6**:eabd4049. DOI: 10.1126/sciadv.abd4049

[25] Villeneuve Paul J, Goldberg MS. Methodological considerations for epidemiological studies of air pollution and the SARS and COVID-19 coronavirus outbreaks. Environmental Health Perspectives. 2020;**128**:095001. DOI: 10.1289/EHP7411

[26] Kogevinas M, Castaño-Vinyals G, Karachaliou M, Espinosa A, Cid Rd, Garcia-Aymerich J, Carreras A, Cortés B, Pleguezuelos V, Jiménez A, et al. Ambient air pollution in relation to SARS-CoV-2 infection, Antibody response, and COVID-19 disease: A cohort study in Catalonia, Spain (COVICAT Study). Environmental Health Perspectives; 2021;**129**:117003. DOI: 10.1289/EHP9726

[27] Hansell AL, Villeneuve PJ. Invited perspective: Ambient air pollution and SARS-CoV-2: Research challenges and public health implications.

Environmental Health Perspectives; 2021;**129**:111303. DOI: 10.1289/EHP10540

[28] Gupta S, Rouse BT, Sarangi PP. Did climate change influence the emergence, transmission, and expression of the COVID-19 pandemic? Frontiers in Medicine (Lausanne). 2021;**8**:769208. DOI: 10.3389/fmed.2021.769208

[29] SanJuan-Reyes S, Gómez-Oliván LM, Islas-Flores H. COVID-19 in the environment. Chemosphere. 2021;**263**:127973. DOI: 10.1016/j.chemosphere.2020.127973

[30] Greenhalgh T, Jimenez JL, Prather KA, Tufekci Z, Fisman D, Schooley R. Ten scientific reasons in support of airborne transmission of SARS-CoV-2. The Lancet. 2021;**397**(10285):1603-1605. DOI: 10.1016/S0140-6736(21)00869-2

[31] Zang S-T, Luan J, Li L, Yu H-X, Wu Q-J, Chang Q, et al. Ambient air pollution and COVID-19 risk: Evidence from 35 observational studies. Environmental Research. 2022;**204**:112065. DOI: 10.1016/j.envres.2021.112065

[32] Marquès M, Domingo JL. Positive association between outdoor air pollution and the incidence and severity of COVID-19. A review of the recent scientific evidences. Environmental Research. 2022;**203**:111930. DOI: 10.1016/j.envres.2021.111930

[33] Herrero M, Thornton P. What can COVID-19 teach us about responding to climate change? The Lancet Planetary Health. 2020;**4**:e174. DOI: 10.1016/S2542-5196(20)30085-1

[34] Yong JC, Choy BKC. Noncompliance with safety guidelines as a free-riding strategy: An evolutionary game-theoretic approach to cooperation during the COVID-19 pandemic. Frontiers in Psychology. 2021;**12**:729. DOI: 10.3389/fpsyg.2021.646892

Chapter 5

# Pathogenicity, Characterisation and Impact of Ambient Bio-Aerosols on the Climatic Processes: With a Special Emphasis on the Indian Subcontinent

*Minati Behera, Jyotishree Nath, Sony Pandey, Ramasamy Boopathy and Trupti Das*


## Abstract

Airborne particulate matter contains biological entities from various anthropogenic/biogenic activities. Within 1 nm–100 μm size, these are carried to long distances through various external agents. Identified as potential pathogens, they bring forth substantial economic losses in many parts of the world. Despite these shortcomings, bio-aerosols play a vital role in cloud condensation, ice nucleation, precipitation and various atmospheric processes affecting the hydrological cycle in general. Furthermore, bio-aerosols play a decisive role in the dispersal of reproductive plant parts and fungal spores, which play important roles in the evolution and sustenance of ecosystems. However, there remains substantial knowledge on air micro-biome with respect to their occurrence, transformation, role in climate change, interaction and impact on living organisms, agriculture and ecosystem. The current COVID-19 pandemic is a wakeup call for retrospective analysis of airborne particles to reduce their emission, transmission and health risk hazards while understanding their impact on various atmospheric processes. This chapter identifies the various types of bio-aerosols and systematically includes their prime role in the climatic processes, pathogenicity to the exposed flora and fauna along with an exclusive interrogation into their types and characterisation over the Indian subcontinent with a hugely diverging population and pollution panorama.


**Keywords:** bio-aerosol, types, characterisation, health impacts, pandemic

## 1. Introduction

Recently, the changes in lifestyle have tremendously changed the surrounding, as the freshwater is wastewater, soil is contaminated with xenobiotic as well as quality of air is deteriorating day by day. The air we breathe not only contains oxygen, nitrogen and other trace gases but also contains fine and ultra-fine particulate matter (PM), which has aerosolised organic and inorganic components in association with live and dead cells of organism that may be microbes, plant parts or skin shedding of animal and human, termed as bio-aerosols. These are released from biosphere to the atmosphere and become a significant component to the already existing myriad air pollutants. Their size varies from 1 nm to 100 μm, whereas they are a diverse group of biological materials like **living and dead organisms** (bacteria, fungus, archaea, protozoa, virus and their by-products), **dispersal units** (fungal spore and pollinium) and **fragments and defecations** (plant debris/leaf litter and brochosomes) [1].

### 1.1 Sources of bio-aerosols

In ambient air, bio-aerosols can exist as particular entities or in aggregates with organic matter, particulate matter, water droplets and chemical constituents of aerosols, which provide an amiable condition for maintaining the metabolism of these organic components that can reproduce even under stressful conditions [2–4]. Hence, for this reason, it is crucial to know the sources of bio-aerosols, their metamorphosis during the course of movement to atmosphere and recycling to biosphere, their impact on public health, agriculture as well as ecosystem including their role in global climate change. The initial step towards becoming airborne is to be aerosolized first from contaminated sites [5]. Bio-aerosols are ubiquitous in nature and present in troposphere at an altitude of 10–20 km and even 20–40 km above sea level within the stratosphere [6]. Major sources of airborne microorganisms are municipal dumping areas, waste streams and discharge points, shabby water-soaked buildings, soil, degenerated and fermented plant and animal parts, sewage sludge dumping sites, animal husbandry, fermentation process, agriculturally active areas, animal houses, breeding farms, feedstuff-factory outlets and various other anthropogenic activities [7–9]. Some bio-aerosols are released to the environment by naturally active processes such as through fungal spores, which are emitted through osmotic stress or floor pressure impact whereas some via passive process, equivalent to thallus debris and dried fungal spores mainly due to wind [1, 10].

## 2. Importance of bio-aerosol and the existing knowledge gap

Bio-aerosols play a significant part in cloud condensation, ice nucleation and precipitation influencing hydrological cycle in addition to local weather patterns. Pratt et al. [11] reported that the ice-crystal residues in cloud condensation nuclei and ice nuclei were of organic origin, which accounted for about 33%. Moreover, bio-aerosols play a vital role in the dispersion of reproductive plant parts and fungal spores. Owing to their small dimension, wind movements transport these through any geo-graphical obstacles and are, therefore, the important aspects of genetic alterations between habitats and geographic changes of biomes. Hence, these are key components within the development, evolution and dynamics of ecosystems. However, a majority of bio-aerosols have a negative impact on agriculture, animals and public health. More importantly, some of these hold the potential to penetrate the deeper components of the respiratory system being of respirable sizes, distinctively 0.003 μm for viruses [12], 0.25–20 μm for bacteria [13], 17–58 μm for plant pollen [14] and 1–30 μm for fungi [15]. In addition, bio-aerosols of 1–5 μm size typically stay in the air, whereas bigger particles get deposited on surfaces sooner [16, 17]. Wet or dry deposition on the earth surface facilitates the removal of bio-aerosols from the atmosphere. Wet deposition courses are the main sink of atmospheric aerosols, while dry deposition has a much smaller global meaningful impact on native air quality and public well-being.

Recently, Haddrell and Thomas [18] have identified that it is time to merge the present and novel methods of multidisciplinary research in atmospheric chemistry, aerobiology and molecular biology to grasp the long-sought mechanisms of bio-aerosol transport and disintegration as this discipline has huge lack of expertise and enough hypothesis. Still, there is very less information on air micro-biome in comparison with their counterparts in water and soil. In addition, the lack of standard methods, environmental guidelines and databases makes it difficult to interpret and compare results. In a widely diversified country like India, there has been a significant lack of knowledge of bio-aerosols with most of the related work done in metropolitan cities, mainly Delhi (mega city), Mumbai, Chennai, Kolkata, Pune and few in central and eastern parts of India [2, 19–42]. Despite being a very important area, no field campaign has been reported yet. The methodology followed for the sampling of bio-aerosols and their identification and quantification has to be improvised over the Indian continent, as there are immense gap areas in comparison with other countries. In addition, an in-depth study is required to know more about the biological entities of the atmosphere to reduce their emission to atmosphere as well as to reduce the health risk hazards. The recent pandemic has been a wakeup call. Rise in epidemics is the most important threat to humankind across the globe since the inception of the civilisation [43]. In late 2019, a novel coronavirus illness, later named as COVID-19 outbreak, was initially announced in Wuhan, China. With the increase in the number of confirmed cases throughout the world, the World Health Organisation (WHO) soon declared it as a global pandemic. Since then, researchers have highlighted different aspects of this situation specifically on the detection, transmission, pathogenicity and epidemiology of the virus (WHO, 2020). COVID-19 attributable to severe acute respiratory syndrome coronavirus 2 (SARS-CoV-2) is thought of being transmitted from human to human [44]; however, the pathways of the airborne transmission of these virus are still unclear. All these potential questions need a serious attention in order to establish an organised approach to deal with such grave challenges in the future.

This chapter includes extensive information pertaining to types of bio-aerosol and their impact on the atmosphere and bio-sphere as a whole. Types and impact, monitoring, sampling methods of bio-aerosol research from the purview of the Indian subcontinent and a brief comparison with international prospectus are the crucial aspects of this critical review, which also include a special section on SARS-COV-2 and its potential transmission pathways within aerosol particles.

## 3. Types and impact of bio-aerosols

### 3.1 Bacteria

Myriad bacterial species are present in indoor and outdoor environments up to 1.5-km altitude of the surface boundary layer and typically up to 12-km altitude within the higher troposphere and even within the stratosphere at the altitudes of 20–41 km above the ocean floor. These bacteria have the potential to adapt to harsh environmental conditions and are one of the most successfully thriving life forms on the Earth [1, 6, 16, 45, 46]. The presence of bacteria in atmosphere is strongly influenced by meteorological conditions, seasonal variability, human activities as well as their natural and anthropogenic sources [2, 8, 20, 23, 36]. Bacteria may exist in the form of individual units or in aggregation with organic structures or could also be dispersed into the air through crops, organismal fragments, soil particles, pollen or spores [45–48].

#### *3.1.1 Impact on atmospheric processes*

Microorganisms play an active role in ice nucleation, cloud condensation, distribution and precipitation in global scale. Besides health hazards, bio-aerosols in the 0.43–0.65 and < 0.43 μm scale ranges are unavoidable due to the cloud condensation nuclei-forming effects of aerosols [20]. Some bacteria possess the distinctive potential of synthesising ice-nucleation-active (INA) proteins at low temperature as well as under stress condition. The synthesised INA proteins, which are anchored and highly conserved in a gene [49], get attached to the outer membrane of the cell wall forming oligomers. Several researchers [50–55] proposed that these INA protein oligomers contribute a template for the formation of ice crystal and concluded that higher temperature is likely to be one of the factors to extend INA proteins in oligomers. In addition, it was reported that cell fragments carrying INA proteins are enough to induce freezing. There have been reports [53, 56–58] that the gram-negative bacteria found in leaf surfaces, such as *Pseudomonas syringae*, *Pseudomonas fluorescence*, *Erwinia herbicola*, *Xanthomas campestris* and *Sphingomonas spp.* play a crucial role in ice nucleation. In addition, Akila et al. [59] suggested that the bacteria present in rainwater are significant candidates for ice nucleation. These microorganisms belong to the known ice-nucleating genera *Pseudomonas, Pantoea* and *Bacillus*. Furthermore, Bowers et al. [60] also suggested that bacteria in atmospheric samples are sometimes present in soil environments or on leaf surfaces.

#### *3.1.2 Impact on flora, fauna and human beings*

Airborne bacteria own deadly impact not only on public well-being but also on agriculture and livestock. *Legionella pneumophila*, a rod-shaped, gram-negative non-spore-forming bacterium, is the leading causative agent of legionellosis in humans and livestock. Inhalation is the primary route of an infection; however, the foremost sources are freshwater, portable water, cooling towers or soil. Moreover, the unprotected livestock or contaminated animal products of India are one of the serious causes for legionellosis [61, 62]. *Bacillus anthracis,* which is well known as anthrax, is a gram-positive, rod-formed and spore-forming bacterium causing bio-terror attack. It can have effect on both human and animal over disclosure to infected livestock or contaminated animal products [63, 64]. However, anthrax has been almost eradicated from Western countries, but still it is a cause of concern in the endemic region of India [65–67]. The inhalation of bacterium *Coxiella burnetii* can cause mild-flu-like condition to severe diseases such as pneumonia or hepatitis leading to increasing death rates [68–70]. Moreover, the most predominantly distributed airborne bacteria belong to genus *Bacillus*, *Acinetobacter, Micrococcus spp.* and *Staphylococcus* [2, 20, 21, 31, 71].

### 3.2 Fungi

Ubiquitous fungi are capable to bring forth serious health hazards [7, 24, 38, 39, 72–77]. They proliferate on moist or hygroscopic materials that may be natural or synthetic in both indoors and outdoors. In addition, numerous groups of fungi represent a significant fraction (by mass and number) of the atmospheric aerosol particles. The aerosolized fungi are components of the fungal bodies that are sufficiently small to develop into airborne particles and mainly contain the spores, hyphae and mycelia. The atmospheric fungi are discharged both actively or passively from their parent bodies. Furthermore, fungal aerosols originate primarily from the fungi rising in plant/tree floor or from the fungi thriving within the soil as well as in human beings.

#### *3.2.1 Impact of fungi on atmospheric processes*

Certain fungal spores possess the potential to induce ice nuclei formation in deep convective clouds at relatively warmer conditions for homogeneous ice nucleation. This process is believed to have an influence on the hydrological cycle [76, 78, 79].

#### *3.2.2 Impact of fungi on flora, fauna and human beings*

The spores can adversely have an effect on the plants, animals and human beings as a result of their ample and widespread dispersal in environment [45, 76, 80, 81]. A lot of fungal genera are the foremost reason behind respiratory ailments primarily causing allergies, asthma and pathogenic infections to the respiratory tract. In addition, they are vital promoters for the degradation of cellulosic and non-cellulosic materials in outdoor atmosphere [82]. However, more than 80 genera of fungi have been reported to cause people susceptible to type 1 allergies [83], among which *Alternaria* is one of the vital allergenic fungi that cause bronchial asthma in chindren [84]. In addition, many common airborne fungal species of *Aspergillus, Penicillium* and *Cladosporium* also cause allergy [85]. The prevalence of childhood asthma is increased in developing countries due to rapid industrialisation [86] as well as distinct seasonal variation, which has resulted in declining air quality. In India, about 15–20 million individuals have bronchial asthma induced by ambient air exposure, according to the World Health Organisation. Apart from that, fungal spores exposure can cause allergic reactions such as allergic sinusitis [87, 88], atopic dermatitis [89] and diseases like sick building syndrome (SBS) [90], and their mycotoxins, especially, can adversely have an effect on human as well as animal well-being. Moreover, most of the fungal species present in ambient air topple within the range of 3.3–2.1 μm. This is in consistent with the extent to which the secondary bronchi of the lungs penetrate the human body. Most of the immuno-toxic and allergic fungi can have an effect on the secondary bronchi in human lungs [91]. Other than that, a lot of fungal plant pathogens trigger considerable losses in agricultural crops in each part of the world, which may be approximately 10,000 different types of fungal-infected plant diseases that may not necessarily be host specific [34].

### 3.3 Viruses

Viruses are the smallest bio-aerosols present in the atmosphere (size approximately 0.003 μm) and are considered as major environmental risk factors to human, animal and plant species [92, 93]. However, their occurrence in the atmosphere is mostly in association with suspended particles [94]. Viruses are dispersed to the atmosphere from water, soil surface and from infected animals, plants, birds and human beings [45]. The detection and identification of these viruses is still a challenge. However, the recent development of molecular tools is proving to be quite promising for their identification. There is little information on viruses in the atmosphere; however, most of the studies are focused on single viral aerosol [93–96].

### *3.3.1 Impact of viruses on atmospheric processes*

Viruses get dispersed easily into the atmosphere pertaining to their sizes. Some virus particles, such as influenza A virus (IAV), upon transmission experience rapid evaporation and shrinkage followed by their encounter with the unsaturated ambient atmosphere. The size of the virus determines the aerodynamic behaviour and its sustainability in the atmosphere. Some settle on the ground quickly with a potential to cause secondary infections. There is little information about ice-nucleating viruses and other environmental factors, such as temperature, humidity and solar radiation, which are affecting the survival of virus in atmosphere [97].

### *3.3.2 Impact of viruses on flora, fauna and human beings*

Airborne viruses are potential environmental risk factors due to their complex disease pathogenesis in humans, plants and animals. Readily transmitted by air, they have a potential to incorporate severe acute respiratory syndrome (SARS). Airborne intestinal virus due to sewage contamination, respiratory syncytial virus (RSV), hantavirus from rodent faeces, varicella-zoster virus, measles, mumps and rubella viruses cause viral diseases in humans and animals [16].

### *3.3.3 Viral aerosol transmission in the recent pandemic*

Viral infection causes severe-pneumonia-like symptoms, such as heavy breathing, coughing, sneezing, which, in turn, help overcome the surface tension of the fluid lining present in the respiratory tract causing virus-loaded aerosol formation [98]. Immediately after the expiration of any virus-loaded droplets from an infected person, the larger droplets (>100 μm diameter) settle down on the nearby surface due to gravity and also remain in the air for few hours [99]. It may be re-suspended in air through various human activities as well as meteorological factors, such as local turbulence, differential adiabatic lapse rate, wind velocity, hence becoming more prone to cause infection [100]. However, the smaller (a few nanometres to 100 μm) remain suspended in the air for longer time due to its circulatory flow, ambient relative humidity and the ability to disperse, thereby travelling to a longer distance based on the expiratory action [101]. These smaller droplets are of great concern as in an outdoor environment; these can travel to some kilometre depending on the wind condition and can penetrate deep into the alveolar sac causing severe health issues. It was evident that larger droplets exhaled from sneezing can travel more than 6 m, while due to coughing, these can travel 2 m and 1 m for breathing [102]. The symptoms of SARS-CoV-2 have been observed to appear within 6 days of infection and reach its peak 4 days later. It is evident that COVID-19 virus remains viable for 4 hours on copper surface whereas for 2–3 days on plastic or steel [103].

There are numerous modes of transmission routes for COVID-19 disease, including airborne transmission, endogenous infection, common vehicle, vector, aerosol transmission [104]. The three major modes of viral transmission [105] are aerosol transmission, droplet transmission and autoinoculation of viable virus from contaminated hands to respiratory tract. Droplet transmission occurs by spraying large or small droplet nuclei directly onto the recipient's eyelids or nasal cavity when an infected person coughs, sneezes or even talks. Droplet, airborne and contact [100] are considered to be the three modes of transmission. Large virus-loaded droplets that are released with sufficient momentum by an infected person reach the respiratory organ of the healthy person through droplet transmission. During coughing, a high intrathoracic pressure of high expiratory airflow of 12 L/s breaks up the mucous into smaller droplets. Physical contact with viral droplets deposited on a surface and transmission to its respiratory organ is known as contact transmission, whereas the inhalation of aerosolized viral droplets by a healthy person is considered airborne mode of transmission [106]. However, the significance of transmitting the disease remains unclear [107]. Samples collected from the air outlet exhaust fan of a COVID-19 positive patient's room have reported to be positive referring to the fact that microscopic virus-loaded aerosols would have been displaced by airflows [108]. Interestingly, there are claims [44] that a person in Mangolia, China, has tested positive when he has passed several times indicating that airborne transmission is credible. The airborne transmission evidence even strengthened up by the WHO statement that in indoor environment, the virus-loaded aerosols can be transmitted up to certain distance and result in an accumulation of infection in a short period.

### *3.3.4 Role of air pollution in COVID-19 transmission*

Though it was found that the spread of COVID-19 virus is mainly by human-to-human transmission, in 55 Italian province capitals, the number of confirmed cases was in the days exceeding the $PM_{10}$ threshold limit of 50 $\mu g/m^3$, hence suggesting the transmission mode to be air-to-human transmission rather than the direct way [109]. It was also noticed that polluted cities with low wind pace have a soaring number of COVID-19 circumstances than the much less polluted cities [109]. It was reported that maximum COVID-19 cases were found in the highly polluted regions of northern Italy (SIMA, 2020). Particulate matter performs a vital function within the transport of viable viral droplets as it comprises solid or liquid particles that act as a carrier and allow the virus to flow in airflow for hours to days. In an analysis done by Setti et al. [110], $PM_{10}$ samples were collected from an industrial site of Bergamo province known to be the epicentre of COVID-19 in Italy. According to Bashir et al. [111], utilising Spearman and Kendall correlation confirmed an optimistic association of $PM_{10}$, $PM_{2.5}$, $SO_2$, $NO_2$ and CO ranges with COVID-19. In a signification study conducted by Wu et al. [112], it was concluded that a rise in 1-$\mu g/m^3$ concentration of $PM_{2.5}$ resulted in a rise in 8% death rate of COVID-19. Similarly, Travaglio et al. [113] concluded the positive correlation between high levels of $SO_2$ and $NO_2$ and COVID-19 cases. However, Dhand et al. [106] explored the fact that by considering some completely different parameters such as air quality index (AQI), four ambient air pollution types such as PM10, PM2.5, $NO_2$ and CO and meteorological parameters such as temperature and sunshine duration, a rise in the concentration of these parameters can set off COVID-19 cases. Similar investigation performed by Xu et al. [114] concluded the fact that the increase in AQI can enhance the spread of COVID-19 under low relative humidity.

Accepting the inter-relationship of confirmed COVID-19 cases and air pollution, many scientists worldwide carried out studies including different perspectives of linkage to this fact. Combining three databases of $NO_2$ concentration, atmospheric situation and confirmed COVID-19 instances from 66 areas of France, Germany, Italian republic and Spain, two $NO_2$ hotspots were recognised, and surprisingly, 78% of fatality was confirmed from these two hotspot areas [115]. However, there is not much exploration on the significance of airborne transmission, which may be due to the fact that it is challenging to directly detect the virus travelling in ambient air as a number of external factors are even associated with the movement of viable particles, such as size, temperature, rainfall, humidity, wind speed, wind direction and flow physics [116].

### *3.3.5 Possible prevention measures*

The fact of COVID-19 virus spreading through air can never be denied. Considering the expiration range of SARS-CoV-2 by an infected person, social distancing and masking by using any type of mask can control the situation to a great extent [104]. Homemade cloth mask should be promoted in the countries with the shortage of mask [117]. However, transmission is not the only factor to be assessed, whereas a risk assessment, including the virus, viability of infection and aerosol formation process, is the need of the hour. The identification of the main pollutant in ambient air causing the spread of COVID-19 virus needs a thorough investigation.

Considering the above fact, a link between COVID-19 transmissions with increasing pollution level is emphasised and needs further speculations. It is obvious from the previous SARS-CoV-1 epidemic episodes that long-run exposure to particulate matter, particularly $PM_{2.5}$ can cause over-expression of alveolar angiotensin-converting enzyme receptor-2 (ACE-2), which on exposure to air pollutants can increase the viral load-depleting ACE-2 receptors making them more prone to viral diseases [118]. The airborne transmission of the SARS-CoV-2 viral aerosol is mostly influenced by disease symptoms as well as meteorological factors. The fatality of the disease lies due to past patient history such as whether he/she has been suffering from any pulmonary disorder and chronic disease such as diabetics, obesity, hypertension, anaemia, which decreases the oxygen-carrying capacity and obstructs its supply to the lungs due to pneumonic alveolar congestion. With chronic exposure to air pollution as reported for the urbanites, COVID-19 lethality increases. Though no vaccine has been developed for SARS-CoV-2 yet, prevention is the only solution. As per WHO guidelines, social distancing (2 m) and wearing of multi-layered face mask are the best-known preventive measures to get rid of being infected from SARS-CoV-2 and its airborne transmission.

## 3.4 Pollen

Pollen is the reproductive unit of plant, which is the largest in size among bio-aerosols; therefore, its residence time in atmosphere is short and is prevalent in near surface of atmosphere. However, the atmospheric convection can up-draft these to high altitudes. Having long residence time, the pollens too have the potential to form ice nuclei in different environments [119]. Most of the pollen grains have hard shell, which protects them from adverse environmental conditions. The environmental elements, such as temperature, humidity, rain and wind, can have an effect on the dispersal of pollen.

### *3.4.1 Impact of pollen on flora, fauna and human beings*

Plant pollen and their fragments are one of the first causes of allergy in city facilities. The size of airborne pollen generally ranges between 10 and 50 μm [120]. The airborne bio-pollutants, such as pollen grains, fungal spores and mud mites, include particular proteins and glycoproteins, which react with the immune entity to trigger allergy. They set off allergic irritation within the nasal, conjunctival and/or bronchial mucosa in addition to bronchial asthma [121, 122]. However, the vast majority of sufferers with allergic ailments possess medical signs within the respiratory tract

[122, 123]. Grass pollens are among well-known bio-aerosols causing respiratory allergy in human beings. It was reported that some of the rice pollen also cause respiratory allergy in the field workers and the people living in the vicinity of the rice field. According to researchers in Israel and Poland, the threshold values for pollen range from 4 to 20 grains/$m^3$ air for grass pollen [1].

### 3.5 Algae

Algae are microscopic or macroscopic in sizes dispersed to atmosphere actively or passively and are found in almost all the environments. In comparison with other bio-aerosols, their number is less in atmosphere. Their presence in atmosphere has been reported in several studies worldwide [124–130]. These are rich in bioactive secondary metabolites, which are beneficial in the developments of new pharmaceutical agents [127, 129].

#### *3.5.1 Impact of algae on flora, fauna and human beings*

Despite their usefulness, microflora cause several adverse health consequences. They are capable of contaminating drinking water, act as allergen to humans [129] and deteriorate old-age architectural heritage [131], and some of the taxa produce toxins posing threat to humans [132]. The genera *Chlorella*, *Scenedesmus*, *Chlorococcum*, *Klebsormidium* (*Hormidium*) and *Lyngbya* are most commonly associated with health results [133]. However, algae and cyanobacteria can make a significant contribution to the total load of suspended air particle making a pleasing contribution when inhaled and causing many environmental problems. In addition, these are everlasting constituents of indoor and outdoor environments [134]. They also cause inflammation of pores and skin, hay fever, allergies, headaches, sinusitis, rhinitis, sclerosis and various respiratory illnesses.

## 4. Indoor versus outdoor environment

The risk of hazards is higher in indoor surroundings than outdoor as folks spend most of their time indoors. In addition, the lack of proper ventilation and dispersal mechanism has posed a high risk to human health [135]. The World Health Organisation has recently reported [136] that the inhabitants of damp or mouldy building both in houses and public buildings are at enhanced threat of experiencing respiratory symptoms, respiratory infections and exacerbation of bronchial asthma. In India, approximately 10 million people suffer from Asthma, whereas 15 million from frequent/intermittent Allergic Rhinitis due to the inhalation of air containing various allergens [137]. The foremost sources of indoor bio-aerosols are home dust, fungal spores and hyphae, pollen grains, shedding of human pores and skin cells and actions such as talking, coughing and sneezing, fragments of insects, meals crumbs, pure fibres. However, outdoor bio-aerosols have always been a source of bio-aerosols in a variety of indoor environments [16, 31]. People belonging to open area were more prone to seasonal attack, whereas those living in congested areas were at high risk to perennial attack [138].

Several studies on monitoring and impact of indoor/outdoor bio-aerosols have been conducted to address health risk issues [1, 16, 20, 75, 76, 96, 135, 139–142]. Bio-aerosols are reported for 5–34% of indoor air pollution [67]. In addition, the global burden of diseases was 2.7%, while 2 million deaths occurred per year due to the indoor bio-aerosols that exceeded the annual mortality attributed to malaria

[43, 143, 144]. One of the best examples of indoor bio-aerosols impact is sick building syndrome (SBS) [31, 123]. Moreover, the bio-terror attack (2001), outbreak of SARS (2003) and H1N1 influenza virus (2009) are some of the life-threatening issues due to bio-aerosols. However, it is estimated that the fraction of outdoor particulate matter (PM) below 2.5 in aerodynamic diameters has attributed to 3.1 million deaths in 2010 worldwide [144]. It has been predicted that in near future, 85% of global population of the developing world, in particular children, would grow up exposed to unsafe air both in indoor and outdoor environments due to the lack of major policy for clean air quality [145, 146]. Morey [147] suggested that concentrations above 1000 CFU/ $m^3$ indicate potential microbial contamination, and further investigation in indoor environment is recommended.

## 5. Bio-aerosol characterisation in India

India with 1.2 billion population (2011, Census) is the second most populated nation on the Earth and likewise accounts for 17% of world's population. Besides, it is a rapidly growing nation and is among the top 10 most industrialised nations of the globe. According to the Central Pollution Control Board (CPCB) of India, there are 17 classes of drastically polluting industries (large and medium scale). Higher population density, rapid urbanisation, industrialisation and improper control policy lead to high environmental burden and a significant health risk. In 2012, the WHO reported that about 7 million folks died prematurely from both indoor and outdoor air pollutions. It accounts for one-eighth of all the premature deaths worldwide [144, 148–150]. In 2013, Global Burden of Disease Study (GBD) concluded that 19% of the world's untimely loss of life occurred in India, while air pollution accounts for 620,000 premature death in each year [151]. These reports signify that air pollution is a global environment burden as well as an important risk factor for morbidity and mortality, especially in the developing countries [143], and is also the world's largest single environmental health hazard [144, 150].

Based on massive developmental activities along with topographical/climatic diversity, the Indian subcontinent has become a breeding ground for quite some complex air pollution issues that is taking a toll on the environment in general, impacting the plant animal and human health in particular.

Biological components in the aerosols are important to understand from several perspectives as these are an integral part of the aerosolisation process apart from having a strong impact on various actively spreading airborne diseases across the country and the world. In spite of the predominance of the biological components in ambient air, bio-aerosol research is still in its infancy in a tropical climatic zone like India. In this study, the authors have made a conscious attempt to compile the entire microbial composition detected in the atmospheric particles across eastern, western, northern and southern parts of India. A vivid description of the bacterial and fungal types and their spatial-temporal variation has been summarised along with the detection techniques. This will be a very first report of its kind from the Indian subcontinent.

### 5.1 Northern India

The northern part of India comprises megacity Delhi, tourist places like Rajasthan, Uttar Pradesh, predominantly biomass-burning states such as Punjab, Chandigarh and Industrial state Haryana as well as the pristine Jammu and Kashmir.

#### *5.1.1 Delhi*

The megacity Delhi is the nationwide capital and second largest metropolitan in India having ~16 million inhabitants (2011 Census, India). The city experiences mainly three seasons: monsoon (subtropical climate, average amount of rainfall of ~611.8 mm), excessive chilly winter (dense foggy climate and low atmospheric boundary layer top conditions) and hot summers (temperature ~ 46–49°C). Occasionally, Delhi also experiences dust storms event from the Thar Desert or the Middle East [2]. According to the recent air pollution monitoring data of 2011 and 2015, Delhi is the second most polluted major city in the world [152]. Therefore, we observed that most of the bio-aerosol-monitoring reports were made from Delhi.

Most of the studies have been reported on viable culturable bacterial and fungal concentration in both the outdoor and indoor environments of the city premises of Delhi. The research group of [41] has been actively monitoring the bio-aerosol concentration in both the indoor and outdoor environments of Jawaharlal Nehru University (JNU) and nearby areas. This research group mainly focuses on the concentration of fungi and bacteria (both gram positive and gram negative). The highest concentrations of both gram-positive and gram-negative bacteria were reported near a garbage dump site where gram-negative bacteria were predominant [41]. However, most of the fungal bio-aerosols identified were of respirable sizes and were associated with immune-toxic diseases and allergies. Higher concentrations of viable airborne microbes were detected in post-monsoon than in monsoon season in indoor sites in comparison with the outdoor ones, which may be due to seasonal rain wash [153]. However, Ghosh et al. [31] reported an enchanting connection between the concentration of fungal spores and microorganism in relation to each environmental and human factor. They analysed ambient ranges of viable bio-aerosol (fungi, gram constructive and destructive bacteria) in the Central Library of JNU (five indoors and one outdoor samples) and suggested that most of the fungal species were *Rhizopu soryzae, Aspergillus nidulans* and *Aspergillus flavus*, whereas predominant bacteria were of *Bacillus* and *Coccus* types based on gram staining and morphology observation. They noticed the highest fungal concentration in indoor surroundings, but bacterial concentration for both gram negative and positive has been the highest within the reading corridor. Furthermore, Kumar et al. [8] reported the abundance of culturable airborne bacteria during monsoon and winter season at the building of the School of Environmental Science in JNU. Their observation revealed that the mean concentrations of bacteria were higher throughout the rainy season than in winters, which deviated from the reports from Delhi as well as from data of temperate as well as tropical climate zones. In addition, the percentage of gram-positive bacteria dominated over gram-negative bacteria during both the seasons, and among gram-positive bacteria, cocci dominated over rod-shaped ones. The ambient level of viable fungal bio-aerosols of a sewage treatment plant (from six different segments) of Vasant Kunj, a posh area in Delhi city, depicted eight genera of fungi, and among them, four, for example *Mucor, Rhizopus, Aspergillus* and *Penicillium,* were found in larger number throughout all the seasons [73]. Moreover, Lal et al. [91] identified and characterised size-segregated bio-aerosols at four different sites in JNU campus, Delhi, during all the four seasons. They suggested the highest bio-aerosol concentration (fungi, gram-positive bacteria and gram-negative bacteria) during post-monsoon and the lowest in monsoon season in all the four sites. Fungal bio-aerosols, such as *Penicillium sp., Alternaria sp.* and *Aspergillus sp.*, were predominant.

Culturable airborne bacterial and fungal concentrations were detected near a busy roadside restaurant cluster in New Delhi. A big variation in bacterial and fungal focus was noticed in numerous seasons starting from $1.7 \times 10^4$–$9.8 \times 10^4$ (averaged $6.3 \times 10^4 \pm 2.6 \times 10^4$ cfu $m^{-3}$) and $3.5 \times 10^2$–$9.5 \times 10^3$ ($3.9 \times 10^3 \pm 3.1 \times 10^3$ cfu $m^{-3}$) cfu $m^{-3}$, respectively. Based mostly on 16S rDNA sequencing, *Bacillus* and *Acinetobacter* were predominant microorganisms, whereas predominant fungal genera were *Aspergillus*, *Cladosporium*, *Alternaria* and *Fusarium*, which are popular for bad health effects resulting in quite a few allergic and pathogenic irritation [21]. In another study, Agarwal et al. [2] reported size-segregated bio-aerosols at Okhla landfill of Delhi, known to be a municipal dumping station, and stated that both bacteria and fungi had major peak in winter than summer. Larger focus of bio-aerosols could be related to low atmospheric boundary layer height and beneficial meteorological situations in New Delhi. However, the low concentration in summer may be due to microbial lethal consequences of critical meteorological situations (high temperature and photovoltaic radiation) that are more pronounced than the release of microbial flux due to the effect of summer photovoltaic underfloor heating. In addition, size distribution evaluation confirmed that microorganisms were largely ample in fine particle sizes, that is <0.43–2.1 μm, but few peaks were additionally seen in dimension ranges between 5.8 and >9.0 μm, whereas fungal spores largely peaked in coarse sizes (2.1–5.8 μm). *Bacillus*, *Staphylococcus*, *Streptococcus*, *Klebseilla* and *Escherichia* genera were predominant bacterial strains. However, a lot of identified fungal spores possess negative health effects resulting in quite a few pathogenic inflammations. In addition, in various places of Delhi, that is at a landfill site, agricultural facilities, highway orientation and restaurant cluster sites report bio-aerosol size disparate monitoring in winter, spring and summer seasons [20]. In spring, bacterial particles were enhanced in size ranging between 5.8 and >9.0 μm, but concentrations were higher during winter. More importantly, ample strains were *Bacillus cereus* (16%), *Bacillus licheniformis* (11%), *Bacillus thuringiensis* (9%), *Micrococcus sp*. (7%) and *Acinetobacter sp.* (9%). The variability in bio-aerosol concentration was reported at three different sites, such as land use configuration commercial complex, dumping site and vegetated ridge with respective microclimatic variation in Urban Delhi. It was observed that the dumping site had the maximum bacterial count, whereas the vegetated ridge site had the highest fungal count [23]. Although spatiotemporal variation was quite distinct, the meteorological parameter independently failed to show a uniform and conclusive relationship. From this study, it was interpreted that the land use pattern and human activity seem to play an important role in determining aerosolized microbial diversity than meteorological variables.

### *5.1.2 Kanpur*

Kanpur is an industrialised metropolis in the state of Uttar Pradesh in India and is positioned with the central part of highly polluted Indo-Gangetic Plain (IGP). Many studies on the physical and chemical characterisation of aerosols and their climate impact have been reported from Kanpur. Viable aerosols and particle concentration in ambient atmosphere of Kanpur were analysed during southwest monsoon and post-monsoon (June-October 2015). Gram-negative bacteria were reported to be predominant over gram-positive bacteria and fungi [154]. However, in a separate study, the same group of researchers reported the maximum concentration of bio-aerosols during wintertime (gram-positive bacteria, gram-negative bacteria and fungi) followed by post-monsoon, monsoon and pre-monsoon during June 2015-May

2016 at the same location. In addition, they reported the correlation between meteorological parameters and bio-aerosol concentration and suggested that ambient temperature affects the bacterial concentration, whereas wet precipitation relates to higher abundance of fungi [155].

The northern part of India is the heavily polluted part where life is quite worse. Still, most of the researchers focus on chemical constituents of particulate matter. However, both the chemical and biological constituents have adverse impact on living beings causing numerous respiratory problems. Many of the researchers represented an excellent correlation between the concentrations of culturable bacteria and fungi with meteorological parameters, and some correlate with organic carbon (OC), elemental carbon (EC) or EC/OC, volatile organic carbon (VOC) and total suspended particulate matter (TSPM) [8, 21, 155]. Moreover, some of them used low-volume handy sampler, single stage and size-segregated impactor to enumerate the culturable bio-aerosol concentration [2, 8, 20, 21]. However, most researchers identified bacterial species based on phenotype characteristics and gram staining [2, 8, 23, 31], whereas very few followed PCR amplification and 16S identification to identify bacteria up to species level [20, 21]. In addition, fungal bio-aerosols were identified based on morphology and lacto-phenol cotton blue staining (microscopic observation). Still, there is a huge gap as culturable bio-aerosols are only a fraction of the bacterial and fungal diversity as most of them were unculturable and have potential impact on living beings [46].

### 5.2 Eastern India

The states Bihar, Jharkhand, West Bengal and Odisha are located in the eastern part of the Indian subcontinent. Kolkata, the capital city of West Bengal, is also one of the largest metropolitan cities in the region, while Bihar and West Bengal are in Indo-Gangetic Plain (IGP); Jharkhand, rich in mineral wealth and dense forest, lies in the Chota Nagpur Plateau. Odisha is also rich in mineral wealth and lies along the Eastern Ghats and Deccan Plateau. Most research activities on bio-aerosols were reported in Kolkata compared to other cities in eastern India.

#### *5.2.1 West Bengal*

Kolkata, the capital of West Bengal, is the third most populous city in India having a population of about ~4.5 million according to a 2011 census. In spite of that, this metropolis is the economic, cultural and academic hub of eastern India and is below traffic congestion, poverty, overpopulation and logistic issues associated with the inhabitant bust [156]. However, Sundarbans mangrove forest, the world's biggest mangrove eco-region on the land-ocean boundary of the Ganges delta, is located at approximately 100 km from Kolkata city [157]. Studies on bio-aerosol from the ambient air and both from outdoor and indoor environments have been reported in many parts of urban and sub-urban Kolkata. Mostly, the studies focused on the identification of fungal species using various sampling techniques. Few reports were on the pollen and bacteria as well; mostly, the effect of meteorological parameters was discussed along with the health impact of the identified microorganisms.

Initial studies on airborne fungal load in agricultural environment throughout threshing operation of paddy and wheat crop within the neighbourhood of Barrackpore, West Bengal, were published way back in 1994 [158]. During wheat threshing, a relatively better frequency of *Alternaria tenuissima* was detected followed by *Drechslera sp.* and *Cladosporium herbarum*, while during the threshing of paddy, a high count of *A. humicola* and *A. tenuissima* seen was detected in Boro variety, but in Aman variety, *Helminthosporium oryzae* was dominated followed by *C. herbarum* and a few unidentified yeasts. However, the frequency of *Aspergillus sp.* was observed before and after threshing but was quite negligible during the threshing period. Adhikari et al. [19] attempted quantitative and qualitative strategies to estimate the extent and kinds of indoor airborne fungal spores in two unique cowsheds: one in a rural region of Konnagar (December 1994-September 1995), and the other in a sub-urban space located at Kestapur (during December 1995-September 1996) of West Bengal. In addition, they attempted to emphasise the fact that these species can cause respiratory allergies in sensitive people. In addition, they monitored the spread of spores in the air. This helps predict the spread of fungi that contaminate milk and dairy products rising out of these cowsheds. Moreover, they attempted for searching out the link between the categories and the amount of fungal spores and culturable moulds produced under completely different meteorological parameters because it is believed that the climate plays a completely crucial feature in fungal growth, sporulation and dispersal. Their investigation revealed that 29 airborne spores were from the rural cowshed, whereas 24 were from the sub-urban area. At every location, a comparatively higher frequency of *Cladosporium sp.*, *Aspergillus/Penicillium* group, *Periconia sp.*, *Nigrospora sp.*, and some unidentified ascospores and basidiospores were reported. However, the occurrences of fungal spore were completely exceptional in several meteorological components in each cowshed.

A comparative research of airborne fungal spores was undertaken in five indoor and outdoor environments in Burdwan, West Bengal, for an interval of 2 years utilising Rotorod samplers and sedimentation plates (culture plate). They revealed the lowest count in summer season and the highest in the wet season. However, *Aspergillus* was fairly predominant in all of the environments investigated [26].

The vertical profile of major airborne pollen and spore concentration at totally different heights in an agricultural farm situated in West Bengal, India, revealed that the frequency of tree pollen confirmed roughly good correlation with the increase in heights, whereas pollen from herb/shrub is dominant at decreasing heights throughout all of the three seasons (winter, summer season and rains). However, the smaller fungal spores were dominant at increasing heights, and larger spores and conidia were more often at decrease levels. In addition, *Aspergilli* group, *Cladosporium* and *Nigrospora* were predominant throughout the investigation, whereas *Aspergillus japonicus* and *Drechslera oryzae* were discovered to be potential inflicting respiratory allergy in agricultural workers. The examination was performed utilising Rotorod samplers [25].

Airborne fungal spore concentration was studied in five indoor environments in Santiniketan, West Bengal, India, for an interval of 2 years utilising the Astir 1-day personal volumetric sampler in addition to a Rotorod sampler and sedimentation plates. *Aspergillus sp.* was the very best aerospora contributor within the environment, followed by *Curvularia* and *Cladosporium*. However, during winter, the focus of *Cladosporium* was highest attributable to the presence of a giant variety of saprophytic kinds [159].

Aeromycoflora studies over a paddy field near Barrackpore, West Bengal, were carried out in two consecutive crop seasons (Rabi) using the gravimetric culture plate exposure technique. *Cladosporium* was dominated at the early stage of seeding, which was declined in the later stage with the onset of summer, while *Penicillium* showed a reverse pattern. However, *Curvularia* showed no seasonal variation and was observed throughout the investigation period. In addition, *Alternaria, Fusarium, Helminthosporium* and *Nigrospora* also observed in air were potent phytopathogenic fungi of which *Alternaria was* predominant. However, a species of Alternaria was detected as an epidemic-causing agent in this variety of rice, which causes leaf injury to the host plant [160]. In another study, Uddin [161] investigated the character and number of pathogenic as well as non-pathogenic aeromycoflora in some of the jute fields close to Barrackpore, West Bengal, India. They revealed that *Penicillium* and *Aspergillus* were probably the most dominant saprophytes adopted by *Curvularia* and *Cladosporium* in the initial season, whereas in 1991, *Aspergillus* and *Curvularia* were essentially the most dominant, followed by *Cephlalosporium*, *Penicillium* and *Cladosporium*, together with *Pullularia* within the mid-season. However, *Macrophomina phaseolina*, a pathogenic fungus, appeared irregularly in the initial season but grew to become extra predominant in later phases within the second season. In addition, different pathogens *Helminthosporium* occurred twice and thrice in the sooner part of the two-crop season, whereas *Alternaria*, *Nigrospora* and *Fusarium* occurred very often.

An aerobiological survey revealed seasonal variation in *Trichoderma harzianum* within the outside surroundings of an agricultural farm and its vicinity within the northern fringe of Kolkata city, India, by an Andersen air sampler. It was noticed that the colony rely was elevated throughout February in around 28 days, and the relative humidity was discovered to be a major ($P < 0.05$) factor predicting the incidence of *T. harzianum* in the air [27]. A scientific quantification of the indoor fungal flora of the Shyambazar Metro-Railway Station, Kolkata, was carried out for an interval of 4 months by utilising gravitational settling methodology *via* Petri dishes with potato dextrose agar (PDA) culture media. In this study, it was revealed that *Aspergillus niger* was probably the most prevalent fungal genera followed by *Aspergillus flavus* and *Penicillium sp.* [162]. The fungal spore concentration within the air of Madhyamgram, a sub-urban space close to Kolkata, was carried out for five consecutive years to find out the effect of various meteorological parameters on the frequency of airborne fungal spores and their well-being on native inhabitants with regard to respiratory allergy. Greater than 50 taxa were detected, out of which 15 were allergenic. In addition, the concentration of spore increased throughout early-winter and rainy season and diminished throughout late-winter and mid-summer. Moreover, a positive correlation was discovered between the respiratory allergy instances and the air-spora concentrations [24]. The fungal spectrum of Konnagar city was evaluated using the Burkard personal sampler, which showed 39 types of fungal spores such as *Cladosporium sp., Aspegilli/Penicilli, Nigrospora sp., Periconiasp., Chaetomium sp., Drechslera sp.* and *Alternaria sp.* in a 1-year calendar [30]. One more study monitored the connection of each day bronchial asthma hospital admission of school-age youngsters and each day concentration of outdoor *Alternaria conidia*, ozone, PM10 and climate parameters within the ambiance of Kolkata, utilising nonparametric generalised additive model [25]. Ghosh et al. [32] evaluated the airborne fungal flora of the Nationwide Library, Kolkata, for an interval of 3 months (February-April, 2010) *via* gravitational settling technique utilising Petri dishes with malt extract agar (MEA) media. *Aspergillus niger*, *A. tenuissima*, *C. herbarum* and *Penicillium sp.* were accounted highest in indoor surroundings, whereas outside atmosphere confirmed clear dominance of *Alternaria alternata*, *Asperillus niger*, *A. tenussima*, *C. herbarum*, *C. cladosporioides*, *Curvularia lunata* and *Fusarium oxysporum*. For the enumeration of bacteria and fungi, a qualitative and quantitative research of indoor air in a hospital at Kalyani, West Bengal, India, was carried out using the settle plate method. The aim of this research was to evaluate microbial inhabitants of indoor air of various wards of the hospital and in several sampling time. The highest microbial inhabitants were recorded during evenings in comparison with the morning hours. *Escherichia coli*, *Staphylococcus aureus*, *Pseudomonas aeruginosa* and *Klebsiella sp.* were predominant among bacteria; however, *Aspergillus sp.*, *Fusariumsp*, *Penicillium sp.* and *Candida sp.* were prevalent among fungi [39]. The aeromycoflora of the Institute of Agriculture library, Visva-Bharati, Santiniketan, India, was monitored with the help of Burkard personal 1-day volumetric sampler to suggest preventive measures towards the collections of library and the library personnel's. Mainly, *Drechslera sp., Aspergilli/ penicillin sp., Bispora sp., Basidiospore sp., Ascospore sp., Nigropora sp., Pericornia sp., Fusarium sp., Cladosporium sp.* and *Trichornis sp*. were monitored during the investigation. In addition, the concentration of fungal spores was maximum in September and minimum in November [163]. The quantitative and qualitative analyses of fungal spores in the atmosphere of a sub-urban area of West Bengal, India, near India-Bangladesh border were carried by Chakraborty et al. [164], by using the Burkard personal volumetric sampler (for over all spore count) and the Andersen two-stage sampler (for viable spore). The fungal spore concentration has shown two peak seasons (February-March and September-October) with the maximum contribution of *Aspergillus-Penicillin* group. In addition, from the antigenic extract of 15 spore members, *Aspergillus fumigatus* showed the highest allergic sensitivity followed by *A. niger.* Kashinath et al. [165] carried out a survey on viable airborne fungal spore in two outdoor environments in Farakka, West Bengal, India, for a period of 2 years, using Andersen two-stage volumetric air sampler. They revealed that monthly total viable fungal spore load (CFU/$m^3$) has shown significant negative correlation with monthly average temperature. However, it was depicted that antigenic extracts from dominant culturable fungi showed high allergic potentiality by skin prick tests (SPTs) among local allergic patients, suggesting health issues in sensitised people.

The bio-aerosol studies in eastern India mainly focused on fungal bio-aerosols and their allergenic impact on workers by the skin prick test; very few reported on pollen and the least number of studies were focused on viable bacterial aerosols. Most of the sampling of bio-aerosols was based on the sedimentation method, the Rotorod sampler and the Andersen two-stage sampler. Still, there is scanty information on the correlation of bio-aerosols with aerosols as well as meteorological parameters. In addition, unculturable bio-aerosols were also not explored.

### 5.3 Western India

The Western India includes states such as Maharashtra, Gujrat and Goa as well as the union territories of Dadra Nagar Haveli and Daman and Diu. In its north, it is bounded by the Thar Desert, in east Vindhya Range and the Arabian Sea in the west. Mumbai, a metropolitan city, is the largest city as well as the financial centre of India. The climate is mostly tropical wet and dry and semi-arid. However, during summer, most of the cities have temperature about 40–42°C and in winter near about 6–7°C.

A seasonal variation of fungal propagules in a fruit marketplace environment in Nagpur (India) was analysed by utilising the Rotorod sampler and the uncovered Petri plate technique to enumerate the quantitative and qualitative estimations of fungus. *Aspergillus* was probably the most often and predominantly detected genus, whereas *Cladosporium*, *Penicillium* and *Alternaria* spores were ample that are generally known as allergenic and pathogenic fungi. The elevated focus of airborne spores was throughout December-January, whereas the utmost focus of *Aspergillus* was during summer season months. The work-related exposure of Indian agricultural workers to airborne microorganisms, dirt and endotoxin, in small agricultural facilities (farms, warehouses and mills) within side **Aurangabad area**, India, was also studied [166]. In two farms, throughout the threshing of maize and pearl millet, there was noticeable escalation in the concentration of airborne microbes by two to four orders of magnitude. However, whilst threshing of maize, probably, most of the microorganisms were thermophilic actinomycetes and mesophilic actinomycetes of the genus *Streptomyces*, whereas in the course of threshing of pearl millet, probably, the most predominant were corynebacteria and gram-negative microorganism. Airborne fungal spore and pollen of **Ismail Ysuf college campus at Jogeswari, Mumbai, India**, were analysed by Kakde [167]. Both the qualitative (Petri plates exposed with media) and quantitative (Rotorod air sampler) estimations of the culturable fungal spore were also analysed. *Aspergillus-Penicillium* types were predominant, while the spores of *Deuteromycotina* followed by *Ascomycotina* and *Basidiomycotina* were the highest during July-September and November-December. In addition, the predominant pollen were *Poaceae*, *Amaranthaceae*, *Tridax*, *Cassia* and *Ricinus*. Airborne bacterial species were detected in six wastewater treatment plants in Mumbai, and the impact of endotoxin on the health of the local population was also analysed. Several clinically important bacterial species were detected in the samples, and it was reported that the workers at the treatment plant are exposed to opportunistic and infectious bacteria [29]. Post-harvest diseases through airborne fungi were investigated in a vegetable market of **Nagpur city**. Some widespread vegetables were screened for the isolation of fungi as marketplace pathogen to test post-harvest diseases. Virtually, 59 fungal spore types and 78 species from 33 genera belonging to completely different groups were reported. Essentially, the most dominant types of fungi were of *Aspergillus* followed by *Cladosporium*, *Penicillium*, *Alternaria*, *Fusarium*, *Curvularia*, *Trichoderma* and *Rhizopus*. However, vital mycotoxin-producing fungi similar to *A. flavus*, *A. fumigatus* and *Fusarium moniliforme* were also removed from the vegetables collected from the market [33]. Karne [34] analysed the pathogenic and allergenic fungal bio-aerosols over Jowar, Wheat and groundnut fields with the help of continuous volumetric Tilak air sampler for three consecutive Rabi seasons. Most of the bio-aerosols reported were allergenic as well as pathogenic. Airborne fungi were identified in the 80-year-old college library in Mumbai using viable volumetric sampler. *Aspergillus spp., Penicillium spp., Alternaria spp., Cladosporium sp., Curvularia sp., Trichodermaspp* and *Chaetomium* were the predominant species [168]. Aeromycological study was carried in the closed indoor environment of various laboratories of **Mahatma Jyotiba Phule Campus, RTM Nagpur University**, by using the culture plate exposure method. Altogether, 3368 fungal colonies in 19 genera and 28 species were reported. Among them, *Ascomycota* contributed to more than half of the colonies, whereas *Oomycota* had least count. However, *Zygomycota* and *Deuteromycota* contributed to moderate count, while no *Basidiomycota* was reported. The prevalence of fungal culture on cellulosic material in the laboratory depends on the changing indoor environment [22]. Air mycoflora study was done in order to estimate the presence of bio-allergens in different locations of STRM University Campus, Nanded, India. The concentrations of fungal spores were observed to be dominant during winter compared to rainy and summer seasons. The most common fungi identified were *Albugo*, *Mucor*, *Rhizopus* and *Aspergillus* [169]. In Mumbai, the diversity of airborne bacterial species was studied using six-stage microbial impactor and bio-sampler to elucidate the role of endotoxin in particulate matter that elicits pro-inflammatory response *ex* *vivo* [28]. Gram-positive and sporogenic bacteria corresponding to Bacillus species dominated the concentration of airborne culturable bacterial. This study suggested the predominance of pathogenic/opportunistic bacteria from human or animal and sporulation present within the ambient air environment. The fungal bio-aerosol concentration across the Deonar landfill website, which is one of the major landfills in Mumbai, was monitored by the impaction method. The maximum concentration of fungal bio-aerosols was reported in monsoon season, while the minimum concentration in winter season. Moreover, environmental factors, such as relative humidity and wind speed, showed positive correlation with culturable aeromycoflora. In addition, the prevalent species reported were *Aspergillus, Penicillium, Alternaria, Curvularia, Trichoderma* and *Rhizopus* [38]. A comparative research of airborne microbial burden within the vicinity of two landfill sites in Mumbai was carried out by Patil and Kakde [37] as municipal landfills are rich in organic matter and, hence, are favourable sites for the growth of microorganisms. The heterotrophic bacterial concentration was maximum in monsoon and minimum during summer season. In addition, *Staphylococci* and *Actinomycetes* concentrations were higher in monsoon and minimum during winter season. Moreover, they reported no bacterial-free month during their investigation.

#### *5.3.1 Rajasthan*

The impact of meteorological factors on the spread of airborne bacteria and fungi at Sardar market, Jodhpur, was studied by Naruka and Gaur [170]. Gram-negative bacilli were predominant among bacteria, whereas *Aspergillus sp*. was dominant among fungi. In addition, both bacteria and fungi showed significant seasonal variations.

In the western part of India, the studies on bio-aerosols were also focused on viable fungal and bacterial bio-aerosols based on the sedimentation method, the Rotorod air sampler and the six-stage microbial sampler. The research group of Gangamma [29] performed many studies on the characteristics of airborne bacterial diversity and pro-inflammatory response of particulate matter. However, the unculturable part of the bio-aerosols was neglected; hence, there is also still a large gap in scientific understanding on bio-aerosols.

### 5.4 Southern India

#### *5.4.1 Visakhapatnam*

Visakhapatnam is the 15th largest city in India and the principal commercial hub and tourist place of Andhra Pradesh [171]. Most of the research studies on air microbiome were on indoor and outdoor environments of different places of Visakhapatnam. Reddy et al. [172] carried out an analysis of air microbiome of a food warehouse using gravity Petri plate method. The maximum dominant fungal aerosols reported in the five different places of the warehouses were *Penicillium, Mucor, Rhizopus, Cladosporium, Aspergillus, Alternaria, Trichoderma, Fusarium, Pseudomonas, Proteus vulgari* and *Enterobacter aerogenes.* They reported that the strains of *Aspergillus* produce aflatoxin of about 1927.3 μg/kg. However, Mohan et al. [173] analysed the indoor microbiological air quality of 30 government schools using Koch's sedimentation method. They interpreted that the sources of bacterial aerosols were mostly from classroom, toilets and canteen, whereas fungal concentration was high in libraries and classrooms. In addition, *Bacillus* species, *Staphylococcus species, Micrococcus species, Pseudomonas species, E. coli* and *Serratia species* were predominant, whereas among fungal species *Aspergillus, Mucor, Rhizopus, Alternaria, Penicillium* and *Cladosporium* were frequent. The indoor and outdoor bacterial concentration of AU High School of Chinna Waltair area of Visakhapatnam was evaluated using the Andersen six-stage cascade sampler [174]. The concentration of bacteria in eosin methylene blue (EMB) plates was the lowest, while that in nutrient agar media in both the indoor and outdoor environments was the highest. However, relatively high frequencies of bacteria were reported in summer season than in rainy and winter seasons. Kumari et al. [174] also investigated the indoor and outdoor environments of two private and two government schools in Chinna Waltair area by using the Andersen six-stage viable impactor. They reported higher bacterial levels in the outdoor environments of all the schools as compared to indoor environments. In addition, bacterial species, such as *Bacillus, Micrococcus* and *Staphylococcus*, were the dominant bacterial genera reported. Bomala et al. [171] evaluated both the indoor and outdoor environments of public places, such as RTC complex, government school and college buildings, by the principle of the Andersen air sampler (five stages) for microbiological contamination in air. They reported that the most significant source of fungal aerosols is outdoor environment as well as the soil, water, plant, etc. Moreover, areas affected by frequently large human traffic were significantly more microbially contaminated compared to the school principal's room and the university laboratories, which were least contaminated as a result of the specific characteristics of the rooms. Recently, Lalitha [175] carried out an analysis of bacterial and fungal load in indoor environment of 25 areas in Visakhapatnam city by the settle plate technique. They interpreted that the areas that are rich in vegetation, such as Arilova and Simhachalam, have less number of bacterial and fungal loads and placed in green zone. However, places such as King George Hospital (KGH), Jagadamba, Daba Gardens and Old town have maximum counts and were in red zone. In addition, gram-positive and spore-forming bacteria ruled over gram-negative bacteria. Moreover, the reported fungal colonies were *Cryptococcus sp.* (13%), *Cladosporium sp.* (11%), *Penicillium sp.* (10%), *Alternaria sp.* (9%) and *Rhodotorula sp.*

#### *5.4.2 Hyderabad*

Hyderabad is the capital of Telangana. In 2013, there was an article in *The Times of India*, making it official that the ambient air of Hyderabad was not suitable for breathe (Andhra Pradesh Pollution Control Board report). Aparna et al. [176] carried out an analysis of fungal and pollen bio-aerosols in the ambient air at various major junctions of greater Hyderabad area by using the Rotorod sampler. Most of the sampling was done in the outer habitats. They reported high concentration of *Alternaria*, *Cladosporium* and *Helminthosporium*. However, Mahatma Gandhi bus station, Charminar, Uppal and Dilsukhnagar contributed high concentration of fungal species, while the lowest value was detected in Kukatpally.

#### *5.4.3 Kerala*

Kerala is a south-western state situated at the Malabar Coast of India, which experiences monsoon (June-November; 21–30°C), winter (December-February; 18–33°C) and summer (March-May; 23–37°C) seasons. Jothish and Nayar [177] analysed the airborne fungal spores focus in indoor and outside environments of a sawmill in Palakkad district utilising the Burkard private slide sampler. They revealed higher spore concentration in the indoor environment than the outdoor environment. The dominated airspora were *Aspergillus/Penicillium, Cladosporium, Nigrospora* and *Ganoderma*, whereas *Aspergillus/Penicillium* was predominant in the indoor environment and *Cladosporium* in the outdoor environment.

The viable bio-aerosols were well studied in the southern part of India; most of the sampling was based on the sedimentation method, the Andersen six-stage cascade sampler, the Rotorod sampler and the Burkard personal slide sampler. The research group of Gunthe also performed extensive studies on fungal and bacterial bio-aerosols, their identification based on 16S rDNA sequencing as well as the seasonal variability of allergenic fungal aerosols and the ice-nucleating bacterial species over southern tropical India.

## 6. Bio-aerosol sampling and characterisation

All kinds of bio-aerosol sampling methods, such as filtration, impaction, impingement, cyclone, gravity sampling and electrostatic precipitation, have been developed for the measurement of bio-aerosols. Different types of filter membrane, such as glass filter, quartz filter and polytetrafluoroethylene filter of different pore sizes, have been widely used for the collection of bio-aerosol samples. In addition, different types of cascade impactors have been implemented successfully for the measurement of size-segregated bio-aerosols [2]. The culture-based methods, such as diverse forms of Polymerase chain response (PCR), actual time PCR (RT-PCR), quantitative PCR (qPCR), denaturing gradient gel electrophoresis-PCR (DGGE-PCR), restriction fragment length polymorphism (RFLP) and 16S rDNA/rRNA sequencing, for the identification of each culturable and non-culturable bacterial and fungal bio-aerosols have put the bio-aerosol analysis a promising path to study the variety of atmospheric microflora. Moreover, ultraviolet aerodynamic particle sizer (UV-APS) [178] and ultraviolet induced fluorescent (UV-LIF) [179] were used for the real-time detection of atmospheric bio-aerosols. Recently, Nasrabadi et al. [180] characterised airborne live and dead bacteria by using LIVE/DEAD BacLight viability kit together with UV-APS.

## 7. Bio-aerosol characterisation in international perspectives

Considering the international perspectives, the researchers worldwide carried out rigorous work on bio-aerosols by using efficient and advanced techniques. However, in India, the published literature suggested that most of the work on bio-aerosols is based on culture-dependent methods [2, 20, 28, 75, 76, 91, 181, 182] to isolate and quantify only viable bacteria and fungus of atmospheric importance, which are a fraction of the total microbial diversity of atmosphere. Moreover, most of the identification of culturable bacterial and fungal aerosols is based on morphology, shape, size, specific selected media, staining (gram staining, endospore staining and lacto-phenol cotton blue staining), biochemical analysis and previously published literature [2, 31]. In addition, few researchers followed DNA extraction, PCR, RFLP and 16S rDNA/rRNA sequencing for the identification of the culturable bacterial and fungal aerosols [2, 20, 76]. However, Gangamma [28] tried to quantify endotoxin concentration (EU/m$^3$) in ambient air of specific site, that is municipal wastewater treatment plant in greater Mumbai. Recently, Valsan et al. [181, 182] have used UV-APS and UV-LIS to characterise biological aerosols in real time in the southern region of India.

## 8. Future direction

In India, there is still a substantial lack of information in the field of bio-aerosols; hence, the following areas could be deeply studied to mitigate the present situation as well as the climate change.

- the diversity of both culturable and non-culturable bacterial and fungal aerosols and also endotoxin in different environment

- suitable sampling method to avoid biased data analysis

- pathways for their transmission and their sustainability in atmosphere.

## 9. Conclusion

In the early nineteenth century, the investigation on bio-aerosols was initiated to grasp the prevalence and dispersion of microorganisms and spores in air, which suggested that these were released from the biosphere to atmosphere. Since then, research on bio-aerosols and their impact on environment have been going on to improve the air quality and, hence, reduce the health risk. Worldwide, various research groups have deeply explored the bio-aerosols, their occurrence, transformation, role in climate change, interaction and impact on living organisms, agriculture and ecosystem from molecular to global scales until now. Recently, in East Asia, a significant number of research papers (on bio-aerosol) have been published. Furthermore, other researchers are also involved in intense work on bio-aerosols to understand the science and their influence with the meteorological parameters to mitigate the present climatic situation. Recent space is the time to merge the prevailing and novel methods of multidisciplinary fields, that is atmospheric chemistry, aerobiology and molecular biology to acknowledge the long-sought mechanisms of bio-aerosol transport and decay. Along with these, bio-aerosols can even be discovered useful as forensic fingerprints.

## Acknowledgements

The authors acknowledge the Director of the CSIR-IMMT and Head of E&S Department, CSIR-IMMT, Bhubaneswar for their expression of approval and support. B.R. and T.D. are grateful to ISRO-GBP (ATCTM & ARFI) for the financial support. J.N. is thankful to DST for the financial support.

## Author details

Minati Behera[1], Jyotishree Nath[1,2], Sony Pandey[1], Ramasamy Boopathy[1] and Trupti Das[1*]

1 Environment and Sustainability Department, CSIR-Institute of Minerals and Materials Technology (CSIR-IMMT), Bhubaneswar, India

2 Academy of Scientific and Innovative Research (AcSIR), CSIR-Institute of Minerals and Materials Technology (CSIR-IMMT), Bhubaneswar, India

*Address all correspondence to: truptidas@immt.res.in and truptiimmt@gmail.com

## References

[1] Fröhlich-Nowoisky J, Kampf CJ, Weber B, Huffman JA, Pöhlker C, Andreae MO, et al. Bioaerosols in the earth system: Climate, health, and ecosystem interactions. Atmospheric Research. 2016;**182**:346-376. DOI: 10.1016/j.atmosres.2016.07.018

[2] Agarwal S, Mandal P, Srivastava A. Quantification and characterization of size-segregated bioaerosols at municipal solid waste dumping site in Delhi. Procedia Environmental Sciences. 2016b;**35**:400-407. DOI: 10.1016/j.proenv.2016.07.021

[3] Bowers RM, McCubbin IB, Hallar AG, Fierer N. Seasonal variability in airborne bacterial communities at a high-elevation site. Atmospheric Environment. 2012;**50**:41-49. DOI: 10.1016/j.atmosenv.2012.01.005

[4] Gołofit-Szymczak M, Górny RL. Bacterial and fungal aerosols in air-conditioned office buildings in Warsaw, Poland—The winter season. International Journal of Occupational Safety and Ergonomics. 2010;**16**:465-476. DOI: 10.1080/10803548.2010.11076861

[5] Kelley ST, Gilbert JA. Studying the microbiology of the indoor environment. Genome Biology. 2013;**14**:1-9. DOI: 10.1186/gb-2013-14-2-202

[6] Fahlgren C, Bratbak G, Sandaa RA, Thyrhaug R, Zweifel UL. Diversity of airborne bacteria in samples collected using different devices for aerosol collection. Aerobiologia (Bologna). 2011;**27**:107-120. DOI: 10.1007/s10453-010-9181-z

[7] Adhikari A, Sen MM, Gupta-BhattacharyaS, ChandaS. Airborne viable, non-viable, and allergenic fungi in a rural agricultural area of India: A 2-year study at five outdoor sampling stations. Science of the Total Environment. 2004;**326**:123-141. DOI: 10.1016/j.scitotenv.2003.12.007

[8] Kumar B, Gupta GP, Singh S, Kulshrestha UC. Study of abundance and characterization of Culturable bioaerosol at Delhi, India. International journal of Environmental engineering and management. 2013;**4**:219-226

[9] Pastuszka JS, Kyaw Tha Paw U, Lis DO, Wlazlo A, Ulfig K. Bacterial and fungal aerosol in indoor environment in upper Silesia, Poland. Atmospheric Environment. 2000;**34**:3833-3842. DOI: 10.1016/S1352-2310(99)00527-0

[10] Elbert W, Taylor PE, Andreae MO, Pöschl U. Contribution of fungi to primary biogenic aerosols in the atmosphere: Wet and dry discharged spores, carbohydrates, and inorganic ions. Atmospheric Chemistry and Physics. 2007;**7**:4569-4588. DOI: 10.5194/acp-7-4569-2007

[11] Pratt KA, Demott PJ, French JR, Wang Z, Westphal DL, Heymsfield AJ, et al. In situ detection of biological particles in cloud ice-crystals. Nature Geoscience. 2009;**2**:398-401. DOI: 10.1038/ngeo521

[12] Taylor EJ. Dorland's Illustrated Medical Dictionary. WB Saunders; 1988

[13] Comrie JD, Thomson WAR. Black's medical dictionary. In: Black's Medical Dictionary. London: Adam and Charles Black. 1951

[14] Stanley RG, Linskens HF. Pollen: Biology Biochemistry Management. Springer Science & Business Media; 2012

[15] Gregory PH. The Microbiology of the Atmosphere. London, United Kingdom. [Google Scholar]: Leonard Hill; 1973

[16] Ghosh B, Lal H, Srivastava A. Review of bioaerosols in indoor environment with special reference to sampling, analysis and control mechanisms. Environment International. 2015;**85**:254-272. DOI: 10.1016/j.envint.2015.09.018

[17] Mohr AJ. Fate and transport of mechanism in air. Man. Environmental Microbiology. 2001

[18] Haddrell AE, Thomas RJ. Aerobiology: Experimental considerations, observations, and future tools. Applied and Environmental Microbiology. 2017;**83**:e00809-17. DOI: 10.1128/AEM.00809-17

[19] Adhikari A, Sen MM, Gupta-Bhattacharya S, Chanda S. Studies on airborne fungal spores from two indoor cowsheds of suburban and rural areas of West Bengal, India. Indoor and Built Environment. 1999;**8**:221-229. DOI: 10.1159/000024645

[20] Agarwal S. Seasonal variability in size-segregated airborne bacterial particles and their characterization at different source-sites. Environmental Science and Pollution Research. 2017;**24**:13519-13527. DOI: 10.1007/s11356-017-8705-2

[21] Agarwal S, Mandal P, Majumdar D, Aggarwal SG, Srivastava A. Characterization of bioaerosols and their relation with OC, EC and carbonyl VOCs at a busy roadside restaurants-cluster in New Delhi. Aerosol and Air Quality Research. 2016a;**16**:3198-3211. DOI: 10.4209/aaqr.2016.01.0043

[22] Ankush K, Mn B, Head A. Comparative studies on indoor Aeromycoflora from the laboratories. International Journal Life Sciences. 2014;**2**:318-324

[23] Balyan P, Student PD, Ghosh C, Das S, Banerjee BD. Spatial variation of biogenic aerosols at different land use configurations in urban Delhi. International Journal of Applied Environmental Sciences. 2017;**12**:731-744

[24] Chakrabarti HS, Das S, Gupta-Bhattacharya S. Outdoor airborne fungal spora load in a suburb of Kolkata, India: Its variation, meteorological determinants and health impact. International Journal of Environmental Health Research. 2012;**22**:37-50. DOI: 10.1080/09603123.2011.588323

[25] Chakraborty P, Chakraborty A, Ghosh D, Mandal J, Biswas S, Mukhopadhyay UK, et al. Effect of airborne Alternaria conidia, ozone exposure, PM10 and weather on emergency visits for asthma in school-age children in Kolkata city, India. Aerobiologia (Bologna). 2014;**30**:137-148. DOI: 10.1007/s10453-013-9312-4

[26] Chakraborty S, Sen SK, Bhattacharya K. Indoor and outdoor aeromycological survey in Burdwan, West Bengal, India. Aerobiologia (Bologna). 2000;**16**:211-219. DOI: 10.1023/A:1007644300296

[27] Das S, Gupta-Bhattacharya S. Trichoderma harzianum: Occurrence in the air and clinical significance. Aerobiologia (Bologna). 2009;**25**:137-145. DOI: 10.1007/s10453-009-9119-5

[28] Gangamma S. Characteristics of airborne bacteria in Mumbai urban environment. Science of the Total Environment. 2014;**488-489**:70-74. DOI: 10.1016/j.scitotenv.2014.04.065

[29] Gangamma S, Patil RS, Mukherji S. Characterization and proinflammatory response of airborne biological particles from wastewater treatment plants. Environmental Science & Technology. 2011;**45**:3282-3287. DOI: 10.1021/es103652z

[30] Ghosal K, Bhattachaya SG. Biomonitoring of fungal spores of a river Bank Suburban City, Konnagar and its link and impact on local people health. Indian Journal of Aerobiology. 2012;**25**(1):1-7

[31] Ghosh B, Lal H, Kushwaha R, Hazarika N, Srivastava A, Jain VK. Estimation of bioaerosol in indoor environment in the university library of Delhi. Sustainable Environment Research. 2013;**23**:199-207

[32] Ghosh D, Dhar P, Chakraborty T, Uddin N, Das AK, North PGS, et al. Study of Aeromycoflora in indoor and outdoor environment of National Library, Kolkata. International Journal of Plant, Animal and Environmental Sciences. 2014;**4**:663-672

[33] Kakde UB, Kakde HU. Incidence of post-harvest disease and airborne fungal spores in a vegetable market. Acta Botanica Croatica. 2012;**71**:147-157. DOI: 10.2478/v10184-011-0059-0

[34] Karne AV. Aeromycological investigations in the ambient air over some crop fields in context to pathogenic and allergenic fungal bioaerosols. Nature, Environment and Pollution Technology. 2013;**12**:695-698

[35] Mehta S, Kambli P, Wani K, Tanavde S, Mirgal S, Kelkar-Mane V, et al. Study of bio-aerosols in a prominent temple in Mumbai City. International Journal of Environmental Studies. 2013;**70**:583-593. DOI: 10.1080/00207233.2013.829323

[36] Pahari AK, Dasgupta D, Patil RS, Mukherji S. Emission of bacterial bioaerosols from a composting facility in Maharashtra, India. Waste Management. 2016;**53**:22-31. DOI: 10.1016/j.wasman.2016.04.027

[37] Patil NS, Kakde UB. Assessment of fungal bioaerosol emission in the vicinity of a landfill site in Mumbai, India. International Journal of Environment and Waste Management. 2017;**20**:75. DOI: 10.1504/ijewm.2017.10006955

[38] Patil NS, Kakde UB. Comparative study of airborne microbial burden in the vicinity of two landfill sites in Mumbai, India. International Journal of Pharma and Bio Sciences. 2017a;**8**:55-63. DOI: 10.22376/ijpbs.2017.8.3.b55-63

[39] Paul D, Biswas K, Sengupta C, Sinha SN. Studies on environmental monitoring of Aeromicroflora in a Hospital at Kalyani, West Bengal, India. Frontiers in Environmental Microbiology. 2015;**1**:47-50. DOI: 10.11648/j.fem.20150103.13

[40] Roy S, Chakraborty A, Maitra S, Bhattacharya K. Monitoring of airborne fungal spore load in relation to meteorological factors, air pollutants and allergic symptoms in Farakka, an unexplored biozone of eastern India. Environmental Monitoring and Assessment. 2017;**189**:370. DOI: 10.1007/s10661-017-6044-x

[41] Srivastava A, Singh M, Jain VK. Identification and characterization of size-segregated bioaerosols at Jawaharlal Nehru University, New Delhi. Natural Hazards. 2012;**60**:485-499. DOI: 10.1007/s11069-011-0022-3

[42] Kakde UB. Study of fungal bioaerosols and microbiological deterioration and degradation of library materials. Journal of Research in Biosciences, Agriculture & Technology. 2015;**3**:393-398. DOI: 10.29369/ijrbat.2015.03.ii.0107

[43] Martin WJ, Glass RI, Balbus JM, Collins FS. A major environmental cause of death. Science (80-.). 2011;**334**:180. DOI: 10.1126/science.1213088

[44] Wang J, Du G. COVID-19 may transmit through aerosol. Irish Journal of Medical Science. 2020;**189**:1143-1144. DOI: 10.1007/s11845-020-02218-2

[45] Després VR, Alex Huffman J, Burrows SM, Hoose C, Safatov AS, Buryak G, et al. Primary biological aerosol particles in the atmosphere: A review. Tellus Series B: Chemical and Physical Meteorology. 2012;**64**:111(D7). DOI: 10.3402/tellusb.v64i0.15598

[46] Serrano-Silva N, Calderón-Ezquerro MC. Metagenomic survey of bacterial diversity in the atmosphere of Mexico City using different sampling methods. Environmental Pollution. 2018;**235**:20-29. DOI: 10.1016/j.envpol.2017.12.035

[47] Bovallius Å, Roffey R, Henningson E. Long-range transmission of Bacteria. Annals of the New York Academy of Sciences. 1980;**353**:186-200. DOI: 10.1111/j.1749-6632.1980.tb18922.x

[48] Lighthart B. The ecology of bacteria in the alfresco atmosphere. FEMS Microbiology Ecology. 1997;**23**:263-274. DOI: 10.1111/j.1574-6941.1997.tb00408.x

[49] Warren G, Wolber P. MicroReview Moiecuiar aspects of microbial ice nucieation. Molecular Microbiology. 1991;**5**:239-243

[50] Garnham CP, Campbell RL, Walker VK, Davies PL. Novel dimeric β-helical model of an ice nucleation protein with bridged active sites. BMC Structural Biology. 2011;**11**:1-12. DOI: 10.1186/1472-6807-11-36

[51] Govindarajan AG, Lindow SE. Size of bacterial ice-nucleation sites measured in situ by radiation inactivation analysis. Proceedings of the National Academy of Sciences. 1988;**85**:1334 LP-1331338. DOI: 10.1073/pnas.85.5.1334

[52] Mueller GM, Wolber PK, Warren GJ. Clustering of ice nucleation protein correlates with ice nucleation activity. Cryobiology. 1990;**27**:416-422. DOI: 10.1016/0011-2240(90)90018-Y

[53] Šantl-Temkiv T, Sahyoun M, Finster K, Hartmann S, Augustin-Bauditz S, Stratmann F, et al. Characterization of airborne ice-nucleation-active bacteria and bacterial fragments. Atmospheric Environment. 2015;**109**:105-117. DOI: 10.1016/j.atmosenv.2015.02.060

[54] Schmid D, Pridmore D, Capitani G, Battistutta R, Neeser J-R, Jann A. Molecular organisation of the ice nucleation protein InaV from Pseudomonas syringae. FEBS Letters. 1997;**414**:590-594. DOI: 10.1016/S0014-5793(97)01079-X

[55] Southworth MW, Wolber PK, Warren GJ. Nonlinear relationship between concentration and activity of a bacterial ice nucleation protein. The Journal of Biological Chemistry. 1988;**263**:15211-15216

[56] Hartmann S, Augustin S, Clauss T, Wex H, Šantl-Temkiv T, Voigtländer J, et al. Immersion freezing of ice nucleation active protein complexes. Atmospheric Chemistry and Physics. 2013;**13**:5751-5766. DOI: 10.5194/acp-13-5751-2013

[57] Maki LR, Willoughby KJ. Bacteria as biogenic sources of freezing nuclei. Journal of Applied Meteorology. 1978;**32**(7):1181-1195. DOI: 10.1175/1520-0450(1978)017<1049:babsof>2.0.co;2

[58] Li J, Lee T-C. Bacterial ice nucleation and its potential application in the food industry. Trends in Food Science

and Technology. 1995;**6**:259-265. DOI: 10.1016/S0924-2244(00)89110-4

[59] Akila M, Priyamvada H, Ravikrishna R, Gunthe SS. Characterization of bacterial diversity and ice-nucleating ability during different monsoon seasons over a southern tropical Indian region. Atmospheric Environment. 2018;**191**:387-394. DOI: 10.1016/j.atmosenv.2018.08.026

[60] Bowers RM, McLetchie S, Knight R, Fierer N. Spatial variability in airborne bacterial communities across land-use types and their relationship to the bacterial communities of potential source environments. The ISME Journal. 2011;**5**:601-612. DOI: 10.1038/ismej.2010.167

[61] Leclerc H, Schwartzbrod L, Dei-Cas E. Microbial agents associated with waterborne diseases. Critical Reviews in Microbiology. 2002;**28**:371-409. DOI: 10.1080/1040-840291046768

[62] Van Leuken JPG, Swart AN, Havelaar AH, Van Pul A, Van der Hoek W, Heederik D. Atmospheric dispersion modelling of bioaerosols that are pathogenic to humans and livestock - a review to inform risk assessment studies. Microbial Risk Analysis. 2016;**1**:19-39. DOI: 10.1016/j.mran.2015.07.002

[63] Anderson PD, Bokor G. Bioterrorism: Pathogens as weapons. Journal of Pharmacy Practice. 2012;**25**:521-529. DOI: 10.1177/0897190012456366

[64] Pohanka M, Skládal P. Bacillus anthracis, Francisella tularensis and Yersinia pestis. The most important bacterial warfare agents - review. Folia Microbiologia (Praha). 2009;**54**:263-272. DOI: 10.1007/s12223-009-0046-1

[65] Humbal C, Gautam S, Trivedi U. A review on recent progress in observations, and health effects of bioaerosols. Environment International. 2018;**118**:189-193. DOI: 10.1016/j.envint.2018.05.053

[66] Kankaria A, Nongkynrih B, Gupta SK. Indoor air pollution in India: Implications on health and its control. Indian Journal of Community Medicine. 2014;**39**:203-207. DOI: 10.4103/0970-0218.143019

[67] Srikanth P, Sudharsanam S, Steinberg R. Bio-aerosols in indoor environment: Composition, health effects and analysis. Indian Journal of Medical Microbiology. 2008;**26**:302-312. DOI: 10.4103/0255-0857.43555

[68] Angelakis E, Raoult D. Q fever. Veterinary Microbiology. 2010;**140**:297-309. DOI: 10.1016/j.vetmic.2009.07.016

[69] Dijkstra F, van der Hoek W, Wijers N, Schimmer B, Rietveld A, Wijkmans CJ, et al. The 2007-2010 Q fever epidemic in the Netherlands: Characteristics of notified acute Q fever patients and the association with dairy goat farming. FEMS Immunology and Medical Microbiology. 2012;**64**:3-12

[70] Parker NR, Barralet JH, Bell AM. Q fever. Lancet. 2006;**367**:679-688

[71] Nayak BK. An Aeromycological Survey of Different Bus Stand Environments in Puducherry District India: International Journal of MediPharm Research. 2015a;**1**:178-183. pp. 178-183

[72] Haleem Khan AA, Mohan Karuppayil S. Fungal pollution of indoor environments and its management. Saudi Journal of Biological Sciences. 2012;**19**:405-426. DOI: 10.1016/j.sjbs.2012.06.002

[73] Maharia S, Srivastava A. Influence of seasonal variation on concentration of fungal bioaerosol at a sewage treatment

plant (STP) in Delhi. Aerobiologia (Bologna). 2015;**31**:249-260. DOI: 10.1007/s10453-014-9361-3

[74] Nayak BK. Studies on Airborne Fungal Spores in the Fish Market Environment of Puducherry City India. International Journal of TechnoChem Research; 012015b. pp. 107-111

[75] Priyamvada H, Akila M, Singh RK, Ravikrishna R, Verma RS, Philip L, et al. Terrestrial macrofungal diversity from the tropical dry evergreen biome of southern India and its potential role in aerobiology. PLoS One. 2017a;**12**:1-21. DOI: 10.1371/journal.pone.0169333

[76] Priyamvada H, Singh RK, Akila M, Ravikrishna R, Verma RS, Gunthe SS. Seasonal variation of the dominant allergenic fungal aerosols - one year study from southern Indian region. Scientific Reports. 2017b;**7**:1-12. DOI: 10.1038/s41598-017-11727-7

[77] Singh AB. Pollen and fungal aeroallergens associated with allergy and asthma in India. Global Journal of Immunology and Allergic Disease. 2014;**2**:19-28. DOI: 10.14205/2310-6980.2014.02.01.4

[78] Haga DI, Burrows SM, Iannone R, Wheeler MJ, Mason RH, Chen J, et al. Ice nucleation by fungal spores from the classes agaricomycetes, ustilaginomycetes, and eurotiomycetes, and the effect on the atmospheric transport of these spores. Atmospheric Chemistry and Physics. 2014;**14**:8611-8630. DOI: 10.5194/acp-14-8611-2014

[79] Morris CE, Sands DC, Bardin M, Jaenicke R, Vogel B, Leyronas C, et al. Microbiology and atmospheric processes: An upcoming era of research on bio-meteorology. Biogeosciences Discussions. 2008;**5**:191-212. DOI: 10.5194/bgd-5-191-2008

[80] Brown JKM, Hovmøll MS. Aerial dispersal of pathogens on the global and continental scales and its impact on plant disease. Science (80-.). 2002;**297**:537-541. DOI: 10.1126/science.1072678

[81] Fisher MC, Henk DA, Briggs CJ, Brownstein JS, Madoff LC, McCraw SL, et al. Emerging fungal threats to animal, plant and ecosystem health. Nature. 2012;**484**:186-194. DOI: 10.1038/nature10947

[82] Ayesha T, Mn B. Diversity of air-borne mycoflora from indoor environment of library. International Journal of Life Sciences 5. 2017;**5**:203-210

[83] Simon-Nobbe B, Denk U, Pöll V, Rid R, Breitenbach M. The spectrum of fungal allergy. International Archives of Allergy and Immunology. 2008;**145**:58-86. DOI: 10.1159/000107578

[84] Knutsen AP, Bush RK, Demain JG, Denning DW, Dixit A, Fairs A, et al. Fungi and allergic lower respiratory tract diseases. The Journal of Allergy and Clinical Immunology. 2012;**129**:280-291. DOI: 10.1016/j.jaci.2011.12.970

[85] Shen, H Der, Tam, MF, Tang, R. Bin, Chou, H. Aspergillus and penicillium allergens: Focus on proteases. Current Allergy and Asthma Reports.2007;7:351-356. DOI:10.1007/s11882-007-0053-8

[86] Concepts C. The asthma epidemic. The New England Journal of Medicine. 2007;**356**:1073-1073. DOI: 10.1056/nejmc063596

[87] Kraemer R, Deloséa N, Ballinari P, Gallati S, Crameri R. Effect of allergic bronchopulmonary aspergillosis on lung function in children with

cystic fibrosis. American Journal of Respiratory and Critical Care Medicine. 2006;**174**:1211-1220. DOI: 10.1164/rccm.200603-423OC

[88] Kurup VP, Shen H. Der Banerjee B. Respiratory fungal allergy. Microbes and Infection. 2000;**2**:1101-1110. DOI:10.1016/S1286-4579(00)01264-8

[89] Akiyama H, Fujii K, Yamasaki O, Oono T, Iwatsuki K. Antibacterial action of several tannins against Staphylococcus aureus. The Journal of Antimicrobial Chemotherapy. 2001;**48**:487-491. DOI: 10.1093/jac/48.4.487

[90] Takigawa T, Wang BL, Sakano N, Wang DH, Ogino K, Kishi R. A longitudinal study of environmental risk factors for subjective symptoms associated with sick building syndrome in new dwellings. Science of the Total Environment. 2009;**407**:5223-5228. DOI: 10.1016/j.scitotenv.2009.06.023

[91] Lal H, Ghosh B, Srivastava A, Srivastava A. Identification and characterization of size-segregated bioaerosols at different sites in Delhi. Aerosol and Air Quality Research. 2017;**17**:1470-1481. DOI: 10.4209/aaqr.2015.05.0331

[92] Ghosal K, Gupta-Bhattacharya S. Current glimpse of airborne allergenic pollen in Indian subcontinent. Acta Agrobotanica. 2015;**68**:349-355. DOI: 10.5586/aa.2015.034

[93] Marwal A, Sahu AK, Gaur RK. First report of airborne begomovirus infection in Melia azedarach (pride of India), an ornamental tree in India. Aerobiologia (Bologna). 2014;**30**:211-215. DOI: 10.1007/s10453-013-9319-x

[94] Yang W, Marr LC. Dynamics of airborne influenza a viruses indoors and dependence on humidity. PLoS One. 2011;**6**: e90972. DOI: 10.1371/journal.pone.0021481

[95] Neira V, Rabinowitz P, Rendahl A, Paccha B, Gibbs SG, Torremorell M. Characterization of viral load, viability and persistence of influenza a virus in air and on surfaces of swine production facilities. PLoS One. 2016;**11**:1-11. DOI: 10.1371/journal.pone.0146616

[96] Pyankov OV, Bodnev SA, Pyankova OG, Agranovski IE. Survival of aerosolized coronavirus in the ambient air. Journal of Aerosol Science. 2018;**115**:158-163. DOI: 10.1016/j.jaerosci.2017.09.009

[97] Prussin AJ, Schwake DO, Lin K, Gallagher DL, Buttling L, Marr LC. Survival of the enveloped virus Phi6 in droplets as a function of relative humidity, absolute humidity, and temperature. Applied and Environmental Microbiology. 2018;**84**:1-10. DOI: 10.1128/AEM.00551-18

[98] Thamboo A, Lea J, Sommer DD, Sowerby L, Abdalkhani A, Diamond C, et al. Clinical evidence based review and recommendations of aerosol generating medical procedures in otolaryngology - Head and neck surgery during the COVID-19 pandemic. Journal of Otolaryngology - Head & Neck Surgery. 2020;**49**:1-14. DOI: 10.1186/s40463-020-00425-6

[99] Morawska L. Droplet fate in indoor environments, or can we prevent the spread of infection? Indoor Air. 2006;**16**:335-347. DOI: 10.1111/j.1600-0668.2006.00432.x

[100] Mittal R, Ni R, Seo JH. The flow physics of COVID-19. Journal of Fluid Mechanics. 2020;**894**:1-14. DOI: 10.1017/jfm.2020.330

[101] Bourouiba L. Turbulent gas clouds and respiratory pathogen emissions:

Potential implications for reducing transmission of COVID-19. JAMA The Journal of the American Medical Association. 2020;**323**:1837-1838. DOI: 10.1001/jama.2020.4756

[102] Xie X, Li Y, Chwang ATY, Ho PL, Seto WH. How far droplets can move in indoor environments - revisiting the Wells evaporation-falling curve. Indoor Air. 2007;**17**:211-225. DOI: 10.1111/j.1600-0668.2007.00469.x

[103] van Doremalen N, Bushmaker T, Morris DH, Holbrook MG, Gamble A, Williamson BN, et al. Aerosol and surface stability of SARS-CoV-2 as compared with SARS-CoV-1. The New England Journal of Medicine. 2020;**382**:1564-1567. DOI: 10.1056/NEJMc2004973

[104] Hsiao TC, Chuang HC, Griffith SM, Chen SJ, Young LH. COVID-19: An aerosol's point of view from expiration to transmission to viral-mechanism. Aerosol and Air Quality Research. 2020;**20**:905-910. DOI: 10.4209/aaqr.2020.04.0154

[105] Tellier R. Aerosol transmission of influenza a virus: A review of new studies. Journal of the Royal Society Interface. 2009;**6**. DOI: 10.1098/rsif.2009.0302.focus

[106] Dhand R, Li J. Coughs and sneezes: Their role in transmission of respiratory viral infections, including SARS-CoV-2. American Journal of Respiratory and Critical Care Medicine. 2020;**202**:651-659. DOI: 10.1164/rccm.202004-1263PP

[107] Eslami H, Jalili M. The role of environmental factors to transmission of SARS-CoV-2 (COVID-19). AMB Express. 2020;**10**:1-8. DOI: 10.1186/s13568-020-01028-0

[108] Ong SWX, Tan YK, Chia PY, Lee TH, Ng OT, Wong MSY, et al. Air, surface environmental, and personal protective equipment contamination by severe acute respiratory syndrome coronavirus 2 (SARS-CoV-2) from a symptomatic patient. JAMA The Journal of the American Medical Association. 2020;**323**:1610-1612. DOI: 10.1001/jama.2020.3227

[109] Coccia M. Factors determining the diffusion of COVID-19 and suggested strategy to prevent future accelerated viral infectivity similar to COVID. Science of the Total Environment. 2020;**729**:138474. DOI: 10.1016/j.scitotenv.2020.138474

[110] Setti L, Passarini F, De Gennaro G, Barbieri P, Perrone MG, Borelli M, et al. SARS-Cov-2RNA found on particulate matter of Bergamo in northern Italy: First evidence. Environmental Research. 2020;**188**:109754. DOI: 10.1016/j.envres.2020.109754

[111] Bashir MF, Bilal MABJ, Komal B, Bashir MA, Farooq TH, Iqbal N, et al. Correlation between environmental pollution indicators and COVID-19 pandemic: A brief study in Californian context. Environmental Research. 2020;**187**:109652. DOI: 10.1016/j.envres.2020.109652

[112] Wu X, Nethery RC, Sabath MB, Braun D, Dominici F. Air pollution and COVID-19 mortality in the United States: Strengths and limitations of an ecological regression analysis. Science Advances. 2020;**6**:eabd4049. DOI: 10.1126/sciadv.abd4049

[113] Travaglio M, Yu Y, Popovic R, Selley L, Leal NS, Martins LM. Links between air pollution and COVID-19 in England. Environmental Pollution. 2021;**268**: R 713-R715. DOI: 10.1016/j.envpol.2020.115859

[114] Xu H, Yan C, Fu Q, Xiao K, Yu Y, Han D, et al. Possible environmental

effects on the spread of COVID-19 in China. Science of the Total Environment. 2020;**731**:139211. DOI: 10.1016/j.scitotenv.2020.139211

[115] Ogen Y. Response to the commentary by Alexandra a. Chudnovsky on 'assessing nitrogen dioxide (NO2) levels as a contributing factor to coronavirus (COVID-19) fatality. Science of the Total Environment. 2020;**740**:139239. DOI: 10.1016/j.scitotenv.2020.139239

[116] Stadnytskyi V, Bax CE, Bax A, Anfinrud P. The airborne lifetime of small speech droplets and their potential importance in SARS-CoV-2 transmission. Proceedings of the National Academy of Sciences. 2020;**117**:11875 LP-11811877. DOI: 10.1073/pnas.2006874117

[117] Jindal SK, Aggarwal AN, Christopher DJ, Dhar R, Jindal A. Face masks - a sustainable measure to mitigate COVID-19. The International Journal of Tuberculosis and Lung Disease. 2020;**24**:645-647. DOI: 10.5588/ijtld.20.0220

[118] Frontera A, Martin C, Vlachos K, Sgubin G. Regional air pollution persistence links to COVID-19 infection zoning. The Journal of Infection. 2020;**81**:318-356. DOI: 10.1016/j.jinf.2020.03.045

[119] Wright TP, Petters MD, Hader JD, Morton T, Holder AL. Minimal cooling rate dependence of ice nuclei activity in the immersion mode. Journal of Geophysical Research – Atmospheres. 2013;**118**:10,535-10,543. DOI: 10.1002/jgrd.50810

[120] Kumar R, Kumar D, Singh K, Mavi A, Kumar M. Identification of airborne pollens in Delhi. Indian Journal of Allergy Asthma and Immunology. 2018;**32**:28. DOI: 10.4103/ijaai.ijaai_35_17

[121] Calderón MA, Casale TB, Togias A, Bousquet J, Durham SR, Demoly P. Allergen-specific immunotherapy for respiratory allergies: From meta-analysis to registration and beyond. The Journal of Allergy and Clinical Immunology. 2011;**127**:30-38. DOI: 10.1016/j.jaci.2010.08.024

[122] Toro AR, Córdova JA, Canales M, Morales SRGE, Mardones PP, Leiva GMA. Trends and threshold exceedances analysis of airborne pollen concentrations in metropolitan Santiago Chile. PLoS One. 2015;**10**:1-21. DOI: 10.1371/journal.pone.0123077

[123] Laumbach RJ, Kipen HM. Bioaerosols and sick building syndrome: Particles, inflammation, and allergy. Current Opinion in Allergy and Clinical Immunology. 2005;**5**:135-139. DOI: 10.1097/01.all.0000162305.05105.d0

[124] Broady PA. Diversity, distribution and dispersal of Antarctic terrestrial algae. Biodiversity and Conservation. 1996;**5**:1307-1335. DOI: 10.1007/BF00051981

[125] Marshall WA. Aerial dispersal of lichen soredia in the maritime Antarctic. The New Phytologist. 1996;**134**:523-530. DOI: 10.1111/j.1469-8137.1996.tb04370.x

[126] Marshall WA, Chalmers MO. Airborne dispersal of antarctic terrestrial algae and cyanobacteria. Ecography (Cop.). 1997;**20**:585-594. DOI: 10.1111/j.1600-0587.1997.tb00427.x

[127] Sahu N, Tangutur AD. Airborne algae: Overview of the current status and its implications on the environment. Aerobiologia (Bologna). 2015;**31**:89-97. DOI: 10.1007/s10453-014-9349-z

[128] Sharma NK, Rai AK, Singh S, Brown RM. Airborne algae: Their present status and relevance. Journal

of Phycology. 2007;**43**:615-627. DOI: 10.1111/j.1529-8817.2007.00373.x

[129] Sharma NK, Singh S, Rai AK. Diversity and seasonal variation of viable algal particles in the atmosphere of a subtropical city in India. Environmental Research. 2006;**102**:252-259. DOI: 10.1016/j.envres.2006.04.003

[130] Tormo R, Recio D, Silva I, Muñoz AF. A quantitative investigation of airborne algae and lichen soredia obtained from pollen traps in south-West Spain. European Journal of Phycology. 2001;**36**:385-390. DOI: 10.1080/09670260110001735538

[131] Wakefield RD. Investigations of decayed sandstone colonised by a species of Trentepohlia. Aerobiologia (Bologna). 1996;**12**:19-25. DOI: 10.1007/BF02248119

[132] Benson JM, Hutt JA, Rein K, Boggs SE, Barr EB, Fleming LE. The toxicity of microcystin LR in mice following 7 days of inhalation exposure. Toxicon. 2005;**45**:691-698. DOI: 10.1016/j.toxicon.2005.01.004

[133] Genitsaris S, Kormas KA, Moustaka-Gouni M. Airborne algae and cyanobacteria: Occurrence and related health effects. Frontiers in Bioscience-Elite. 2011;**3 E**:772-787. DOI: 10.2741/e285

[134] Tiberg E, Bergman B, Wictorin B, Willen T. Occurrence of microalgae in indoor and outdoor environments in Sweden. Nord Aerobiology. 1983:24-29

[135] Blais-Lecours P, Perrott P, Duchaine C. Non-culturable bioaerosols in indoor settings: Impact on health and molecular approaches for detection. Atmospheric Environment. 2015;**110**:45-53. DOI: 10.1016/j.atmosenv.2015.03.039

[136] Singh U, Garg A, Rani B, Maheshwari R, Prasad M. Indoor air Pollution & its Impact on public health : A review. Advances in Bioresearch. 2012;**3**:21-26

[137] Mouli PC, Mohan SV, Reddy SJ. Assessment of microbial (bacteria) concentrations of ambient air at semi-arid urban region: Influence of meteorological factors. Applied Ecology and Environmental Research. 2005;**3**:139-149. DOI: 10.15666/aeer/0302_139149

[138] Khan AH, Tiwari A. Airborn Bacteria and Fungi Level in. 2015;**3**:108-118

[139] Liu H, Zhang Z, Wen N, Wang C. Determination and risk assessment of airborne endotoxin concentrations in a university campus. Journal of Aerosol Science. 2018;**115**:146-157. DOI: 10.1016/j.jaerosci.2017.09.002

[140] Mahapatra PS, Jain S, Shrestha S, Senapati S, Puppala SP. Ambient endotoxin in PM10 and association with inflammatory activity, air pollutants, and meteorology, in Chitwan, Nepal. Science of the Total Environment. 2018;**618**:1331-1342. DOI: 10.1016/j.scitotenv.2017.09.249

[141] May S, Romberger DJ, Poole JA. Respiratory health effects of large animal farming environments. Journal of Toxicology and Environmental Health: Part B, Critical Reviews. 2012;**15**:524-541. DOI: 10.1080/10937404.2012.744288

[142] Xie Z, Li Y, Lu R, Li W, Fan C, Liu P, et al. Characteristics of total airborne microbes at various air quality levels. Journal of Aerosol Science. 2018;**116**:57-65. DOI: 10.1016/j.jaerosci.2017.11.001

[143] Lim SS, Vos T, Flaxman AD, Danaei G, Shibuya K, Adair-Rohani H, et al. A comparative risk assessment of burden of disease and injury attributable to 67 risk factors and risk factor clusters in 21 regions, 1990-2010: A systematic

analysis for the global burden of disease study 2010. Lancet. 2012;**380**:2224-2260. DOI: 10.1016/S0140-6736(12)61766-8

[144] WHO. [cited 2013 16 November]. Available from: http://www.who.int/indoorair/en/index.html

[145] Brugha R, Grigg J. Urban air pollution and respiratory infections. Paediatric Respiratory Reviews. 2014;**15**:194-199. DOI: 10.1016/j.prrv.2014.03.001

[146] Global Atmosphere Watch. Wmo/igacimpacts of megacities on air pollution and climate. In: Gaw Report no. 205. Geneva, Switzerland: World Meteorological Organization; 2012

[147] Morey PR. Case presentations: Problems caused by moisture in occupied spaces of office buildings. In: Ann. Am. Conf. Gov. Ind. Hyg. 1984. pp. 121-127

[148] Gurjar BR, Ravindra K, Nagpure AS. Air pollution trends over Indian megacities and their local-to-global implications. Atmospheric Environment. 2016;**142**:475-495. DOI: 10.1016/j.atmosenv.2016.06.030

[149] Pant P, Guttikunda SK, Peltier RE. Exposure to particulate matter in India: A synthesis of findings and future directions. Environmental Research. 2016;**147**:480-496. DOI: 10.1016/j.envres.2016.03.011

[150] Ravindra K, Sidhu MK, Mor S, John S, Pyne S. Air pollution in India: Bridging the gap between science and policy. Journal of Hazardous, Toxic & Radioactive Waste. 2016;**20**:A4015003. DOI: 10.1061/(asce)hz.2153-5515.0000303

[151] Feigin VL, Roth GA, Naghavi M, Parmar P, Krishnamurthi R, Chugh S, et al. Global burden of stroke and risk factors in 188 countries, during 1990-2013: A systematic analysis for the global burden of disease study 2013. Lancet Neurology. 2016;**15**:913-924. DOI: 10.1016/S1474-4422(16)30073-4

[152] WHO. Global Urban Ambient Air Pollution Database (Update 2016). Available at: http://www.who.int/phe/health_topics/outdoorair/databases/cities/en/. [Accessed on 30 July 2017]

[153] Lal H, Punia T, Ghosh B, Srivastava A, Jain VK. Comparative study of bioaerosol during monsoon and post-monsoon seasons At four sensitive sites in Delhi. International Journal of Advanced Earth Environmental Science Comp. 2013;**1**:1-7

[154] Rajput P, Gupta T. Bioaerosols & Ambient Meteorology over Kanpur in Northern India Bioaerosols & Ambient Meteorology over Kanpur in Northern India. Europe: International conference on modelling of environmental and water resources; 2017

[155] Rajput P, Anjum MH, Gupta T. One year record of bioaerosols and particles concentration in indo-Gangetic plain: Implications of biomass burning emissions to high-level of endotoxin exposure. Environmental Pollution. 2017;**224**:98-106. DOI: 10.1016/j.envpol.2017.01.045

[156] Das R, Khezri B, Srivastava B, Datta S, Sikdar PK, Webster RD, et al. Trace element composition of PM2.5 and PM10 from Kolkata–a heavily polluted indian metropolis. Atmospheric Pollution Research. 2015;**6**:742-750. DOI: 10.5094/APR.2015.083

[157] Chatterjee A, Sarkar C, Adak A, Mukherjee U, Ghosh SK, Raha S. Ambient air quality during diwali festival over Kolkata - a mega-city in India. Aerosol and Air Quality Research.

2013;**13**:1133-1144. DOI: 10.4209/aaqr.2012.03.0062

[158] Uddin N, Chakraverty R. Airborne fungal load in agricultural environment during threshing operations. Mycopathologia. 1994;**127**:145-149. DOI: 10.1007/BF01102914

[159] Bhattacharya K, Raha S, Majumdar MR. Measuring indoor fungal contaminants in rural West Bengal, India, with reference to allergy symptoms. Indoor and Built Environment. 2001;**10**:40-47. DOI: 10.1159/000049211

[160] Uddin N. Airspora studies over a rice (high yielding variety) field in rabi season in the state of West Bengal, India. Aerobiologia (Bologna). 2004;**20**:127-134. DOI: 10.1023/B:AERO.0000032946.94242.52

[161] Uddin N. Estimation of aeromycoflora in jute fields. Aerobiologia (Bologna). 2005;**21**:75-80. DOI: 10.1007/s10453-004-5883-4

[162] Ghosh D. Identification and distribution of aeromycoflora in the indoor environment of Shyambazar metro-Railway Station, Kolkata, India. African Journal of Microbiology Research. 2011;**5**:5569-5574. DOI: 10.5897/ajmr10.765

[163] Ray PP. Indoor Aeromycroflora at Institute of Agriculture Library (Visva-Bharati): A study. SRELS Journal of information Management. 2017;**54**:37. DOI: 10.17821/srels/2017/v54i1/92575

[164] Chakraborty P, Ghosal K, Karak P, Sarkar E. Airborne Fungal Spores in a Suburban Area of Eastern India with Reference to their Allergenic Potential and Effect on Asthma. India: Journal of Palynology; 2018

[165] Kashinath B, Roy S, Chakraborty A. Analysis of Culturable Aeromycoflora in Farakka, West Bengal, India in Relation to Meteorological Parameters, Air Pollutants and Allergic Manifestation. India: Today and tomorrow's Printers and Publishers; 2018

[166] Krysińska-Traczyk E, Pande BN, Skórska C, Sitkowska J, Prazmo Z, Cholewa G, et al. Exposure of Indian agricultural workers to airborne microorganisms, dust and endotoxin during handling of various plant products. Annals of Agricultural and Environmental Medicine. 2005;**12**:269-275

[167] Kakde U. Exposure to bioaerosols: Survey of bioaerosol in different indoor and outdoor working environments. Bionano Frontier. 2009;**2**:58-60

[168] Kakde UB. Fungal bioaerosols: Global diversity, distribution and its impact on human beings and agricultural crops. Bionano Frontier. 2012;**5**:323-329

[169] Yannawar DV. Determination of Culturable indoor airborne Fungi In campus areas of Srtm university Nanded, India. Journal of Applied Sciences in Environmental Sanitation. 2014;**9**:74-78

[170] Naruka K, Gaur J. Distribution pattern of airborne Bacteria and Fungi at market area. American Journal of Scientific Research. 2014;**9**(9):186-192. DOI: 10.5829/idosi.aejsr.2014.9.6.86254

[171] Bomala K, Saramanda G, Reddy B, Kaparapu J. Microbiological indoor and outdoor air quality of selected places in Visakhapatnam City, India. International Journal of Current Research. 2016;**5**:29059-29062

[172] Reddy MK, Srinivas T, Lakshmi KA. A study of bioaerosols in indoor air of food Godowns of Visakhapatnam, India. Journal of Environmental Research And Development. 2012;**6**:446-451

[173] Mohan KNM, Ramprasad S, Maruthi YA. Microbiological air quality of indoors in primary and secondary schools of Visakhapatnam, India. International Journal of Current Microbiology and Applied Sciences. 2014;**3**:880-887

[174] Kumari KN, Shravanthi CM, Reddy TB. Identification and assessment of airborne bacteria in selected school environments in Visakhapatnam, India. Indian Journal of Science and Technology. 2015;**3**:21-25

[175] Lalitha C. Contamination of refrigerator is a threat for infections. International Journal of Advanced Research. 2019;**5**:1514-1517

[176] Aparna L, Mythili K, Ratnam V, Uram S, Technological S, Technological S, et al. Subject : Environmental Science IJRIES. 2020:1-14

[177] Jothish PS, Nayar TS. Airborne fungal spores in a sawmill environment in Palakkad District, Kerala, India. Aerobiologia (Bologna). 2004;**20**:75-81. DOI: 10.1023/B:AERO.0000022981.70984.b7

[178] Huffman JA, Prenni AJ, DeMott PJ, Pöhlker C, Mason RH, Robinson NH, et al. High concentrations of biological aerosol particles and ice nuclei during and after rain. Atmospheric Chemistry and Physics. 2013;**13**:6151

[179] Gosselin MI, Rathnayake CM, Crawford I, Pöhlker C, Fröhlich-Nowoisky J, Schmer B, et al. Fluorescent bioaerosol particle, molecular tracer, and fungal spore concentrations during dry and rainy periods in a semi-arid forest. Atmospheric Chemistry and Physics. 2016;**16**(23)

[180] Nasrabadi AM, An S, Kwon S-B, Hwang J. Investigation of live and dead status of airborne bacteria using UVAPS with LIVE/DEAD® BacLight kit. Journal of Aerosol Science. 2018;**115**:181-189. DOI: 10.1016/j.jaerosci.2017.10.012

[181] Valsan AE, Priyamvada H, Ravikrishna R, Després VR, Biju CV, Sahu LK, et al. Morphological characteristics of bioaerosols from contrasting locations in southern tropical India - a case study. Atmospheric Environment. 2015;**122**:321-331. DOI: 10.1016/j.atmosenv.2015.09.071

[182] Valsan AE, Ravikrishna R, Biju CV, Pöhlker C, Després VR, Huffman JA, et al. Fluorescent biological aerosol particle measurements at a tropical high-altitude site in southern India during the southwest monsoon season. Atmospheric Chemistry and Physics. 2016;**16**:9805-9830. DOI: 10.5194/acp-16-9805-2016

Section 3

# Protection from Air Pollution

Chapter 6

# Impacts of the Indoor Air Quality on the Health of the Employee and Protection against These Impacts

*Ferdi Tanir and Burak Mete*


## Abstract

Workplace indoor factors are among the factor which affect the health of workers most in all sectors. Another important one of these factors is the air quality of the workplace. There are three main groups of workplace indoor pollutants: biological ones, chemical ones and particles (non-biological). They were grouped as asbestos, biological pollutants, carbon monoxide (CO), carbon dioxide ($CO_2$), formaldehyde/pressed wooden products, lead (Pb), nitrogen dioxide, ($NO_2$), Radon (Rn), indoor particle substances, environmental tobacco smoke, volatile organic compounds (VOCs), humidity, odor and wood smoke. The workplace indoor air pollutants are among the primary indoor air pollutants with serious effects on health and the potential to deteriorate the workers' health. Healthy indoor air quality is defined as the indoor air which does not contain hazardous substances and does not create sense of discomfort in at least 80% of the workers in the workplace. Poor indoor may result in a variety of health problems, from headache, dizziness and nausea, to asthma, cardiovascular diseases, cancer and death.


**Keywords:** air quality, worker, health effect, prevention, VOCs

## 1. Introduction

Oxygen is a basic need for the survival of human beings who can stand its deficiency only for a certain period of time which is shorter when compared with their other needs. Human beings receive oxygen from air through respiration. It is a right for them to have clean air around them. Pollution of air have adverse effects on the health of living things. The relation between health and air quality has been a fact acknowledged since Hippocrates. Indoor environments may pollute the air with the outdoor pollutants and indoor pollutants. Indoor air quality is of particular importance for people who spend 90% of their time in public or private indoors such as houses, schools, fitness centers, shopping malls, supermarkets, workplaces and transportation vehicles. The health risks accompanying exposure to indoor air pollution can be worse for many people when compared with those of the outdoor air pollution.

Unhealthy indoor air may be harmful particularly for risk groups, such as children, the elderly or the patients suffering chronic disorders. Workplace indoor air quality is equally important for those spending at least 1/3 of their lifetime in workplace [1, 2].

Workplace indoor factors are among the factor which affect the health of workers most in all sectors. Another important one of these factors is the air quality of the workplace. One of the focal points of occupational health is environmental effect on the workers working in the indoors of workplaces, when compared to those working in industrial workplaces, such as construction, mining and agricultural workplaces which are very dangerous or dangerous. Furthermore, workers working indoors are less prepared or experienced against environment risks in comparison to the ones working in industrial workplaces. Overview of relevant legislation shows control mechanisms used in such workplaces are not sufficient [3]. Workplace indoor air quality is an optimal indoor requirement which ensures health, comfort and wellbeing of workers and includes minimum air pollutants. Indoor air quality varies according to air temperature, relative humidity, air speed and chemicals at workplace [4]. In today's workplaces, materials, equipment, various cleaning products and chemical and particle emissions determine the indoor air quality. Indoor air quality affect workers' health, incidence of occupational accidents, nonattendance and productivity. For this reason, ensuring that the workplace indoor air quality conforms to the norms is of particular importance. Indoor air quality affects workers' health, emergence of occupational accidents, in attendance and productivity [5]. This article aims at shedding light on the adverse effects of workplace indoor air quality on the health of workers and also the measures which need to be taken for preventing such effects. For this reason, this part covers information on factors deteriorating indoor air pollution, health problems which might develop in workers as a result of the indoor air quality, workplace indoor air quality assessment criteria stemming from international and national source data, the measures to be taken to protect workers from indoor air pollution, indoor air quality and Covid-19.

## 2. Factors affecting workplace indoor air quality

**Indoor air pollutants originating from environment/outdoor environment:** These are biological factors, industrial pollutants, fire products, ammonia, ozone, traffic pollutants such as nitrogen dioxide and particles, radon, methane and humidity.

**Indoor air pollutants originating from workplace:** They were grouped as asbestos, biological pollutants, carbon monoxide (CO), carbon dioxide ($CO_2$), formaldehyde/pressed wooden products, lead (Pb), nitrogen dioxide, ($NO_2$), Radon (Rn), indoor particle substances, environmental tobacco smoke, volatile organic compounds (VOCs), humidity, odor and wood smoke [6–8]. Workers working in the workplace indoor environment may be exposed to various air pollutants (both in gas form and particle form), including organic, inorganic and biological ones. The workplace indoor air pollutants are among the primary indoor air pollutants with serious effects on health and the potential to deteriorate the workers' health. Indoor air quality is affected by the tobacco and nicotine products, chemicals used for cleaning purposes, heating, construction materials and humidity. The pollutants in the indoor air can be present in the form of gas or particle and they may cause various diseases in the respiration system. The most important indoor air pollutant is the smoke of cigarette. The smoke produced by smokers includes many hazardous substances, just like the smoldering cigarette. Most of them are carcinogenic. The people who are exposed to this smoke may develop serious diseases, although they do not consume these products. All workers are affected by the smoke of cigarette consumed in the environment. Limiting the act of smoking to particular rooms would not protect other workers. The particles of tobacco products hang in the air for five hours. Hookah, warmed up tobacco products and electronic cigarette consumption have the same effects on environment and people [6–8]. The construction materials, equipment and chemicals present in the workplace deteriorates the indoor air quality. There are sources of pollutants which deteriorates the indoor air quality and health of workers can be protected only if their effects are eliminated. There are three main groups of workplace indoor pollutants: biological ones, chemical ones and particles (non-biological):

**Biological factors:** They may stem from excessive bacteria, virus, fungus, dust mite, animal hair, pollen concentrations, insufficient maintenance and cleaning, water split, insufficient humidity control, condensation or leakages at the building envelope or water leakage caused by flood.

**Chemical Factors:** Chemical pollutants (gas and vapor), emissions stemming from the products used in the building (like floor or walls covering, office equipment, furniture, insecticide, cleaning products), accidental spillage of chemicals and products used for construction purposes, adhesives, paints and combustion products such as carbon monoxide, formaldehyde and nitrogen dioxide are included in this group.

**Particle-associated factors (non-biological):** Particles are substances which are light enough to hang in the air, in solid or liquid form, and they are non-biological. The building may extract dust, dirt or other substances. Particles may be produced by activities such as construction, wooden punching, drywall, printing, duplication, copying and operation equipment [9].

## 3. Health problems associated with workplace indoor air pollution

Healthy indoor air quality (IAQ) is defined as the indoor air which does not contain hazardous substances and does not create sense of discomfort in at least 80% of the workers in the workplace [8–10]. Poor indoor may result in a variety of health problems, from headache, dizziness and nausea, to asthma, cardiovascular diseases, cancer and death. The typical effect of some common indoor air pollutants on the health and wellbeing of the residents can be seen in **Table 1**. Poor air quality may also have adverse impact on workplace performance, learning at the education/training institutions and improvement of health services, in addition to being hazardous in terms of health and comfort. The impact of indoor air pollutants on the health of the people depends on the concentration of the concerned pollutant, exposure duration, age and gender of the people exposed to it. As for the industrial workplaces, most of the primary risks are evaluated in terms of use of personal protection equipment (PPE), exposure risk and local air conditioning [7–10].

EU-OFFICER research Project showed the association between the indoor chemicals in the office and sick building syndrome-SBS symptoms. The most expressed complaints are ocular irritation (dry eyes, watering eyes or itching, burning or irritation), headache, lethargy, extraordinary tiredness. The researchers also reported that xylene, ethylbenzene, α-pinene, d-limonene, styrene, formaldehyde, acrolein, propionaldehyde, hexane and ozone might increase in the incidence of the symptoms. They concluded by underlining the need for further research in order to better depict

| Pollutant | Impact on health |
|---|---|
| **Carbon monoxide (CO)** | Carbon monoxide can cause headaches, dizziness, nausea and at very high levels, death. Elderly people, pregnant women, young children and people with heart disease and lung disease are more sensitive to the adverse effects of carbon monoxide. |
| **Formaldehyde** | Formaldehyde can cause eye, nose and throat irritation and is considered a potential human carcinogen. |
| **Nitrogen dioxide ($NO_2$)** | Exposure to nitrogen dioxide can cause inflammation of the airways, respiratory illnesses and possibly increases the risk of lung infections. Young children and people with asthma are the most sensitive to $NO_2$. It plays a major role in the development of chronic obstructive pulmonary disease in adults which will affect more people than heart disease by 2020 (Environmentalist 2012). Long-term exposure may also affect lung function and can enhance responses to allergens in sensitized individuals. |
| **Odor** | Odorous discharges are subjective and cause nausea and irritation for some people. |
| **Ozone ($O_3$)** | Ozone exposure can cause asthma, irritation and damage to the eyes, nose and airways. Prolonged exposure to high levels may result in damage to the lungs and airway linings. |
| **Particulate matter** | Inhalable particles have been linked with a number of respiratory illnesses, including asthma and chronic bronchitis. Long-term exposure to fine particles can cause premature death from heart disease and lung disease including cancer. Short-term exposure to higher levels of fine particle concentrations have also been linked with cardio-vascular problems and increased death rates. Exposure to fine particles has also been linked to prevalent anxiety and hypertensive disorders. |
| **Volatile organic Compounds (VOCs)** | Key symptoms associated with exposure to VOCs include eye irritation, nose and throat discomfort, headache and allergic skin reaction. |

**Table 1.**
*Typical health impacts of some common pollutants found indoors.*

the complicated relationship between IAQ and health interaction symptoms [11]. A research conducted on the effects of indoor air quality on the health of workers in Middle East showed that the first most affected part of workers' bodies is their respiratory system; the second most affected one is their cardiovascular system, and the third most affected one is their visual system [12].

## 4. Indoor air pollution for health professionals

People living in developed European and American countries are reported to be more exposed to airborne substances deriving from indoors where they spend most of their time (>90%) causing environmentally associated symptoms that should be evaluated by health professionals. However, this percentage is expected to be higher for infants and the elderly, chronically ill people and in urban settings [13]. Many pollutants present with higher concentrations indoor than outdoor, especially in case of longer and non-intermittent exposure like in the home, workplace and school. It should be taken into account that some of the signs and symptoms presented in the text may occur only in the case of significant exposures. However, lower or shorter exposures with milder or indeterminate symptoms, or atypical presentation (noted in the text) in younger aged children render the diagnosis more difficult. The cooperation of the individual and the health care professional is essential for the correct diagnosis noting clues suggestive of indoor air pollution, like time patterns or location of occurrence by the help of a log or diary of symptoms. In the absence of this cooperation, the following questions in addition to the medical history may be useful.

- Start, duration and periodicity (diurnal, daily, weekly, seasonally) of symptom or complaints.
- Their relation with location under consideration (cessation when away or reoccurrence when returning)
- The work type, work place, any change including moving or decoration.
- Exposure to environmental tobacco smoke at work, school, home, etc.
- Place of residence including internal change or moving
- Exposure to a new hobby etc., a new pet.
- Similar problem in anybody in close contact at home or work (**Table 2**).

### 4.1 Health problems related to environmental tobacco smoke (ETS)

It is the most observed indoor air contaminants. It easily disperses and it is hard not be inhaled by workers in the workplace.

**Key Signs/Symptoms in Adults;** conjunctival irritation, headache, persistent cough, wheezing, rhinitis/pharyngitis, nasal congestion, exacerbation of chronic respiratory conditions.

**Key Signs/Symptoms in Infants and Children;** asthma onset, snoring, bronchitis, repeated pneumonia, persistent middle-ear effusion, frequent upper respiratory infections and/or episodes of otitis media, increased severity of, or difficulty in controlling, asthma.

ETS is also defined as Group A human carcinogen by the U.S. Environmental Protection Agency (EPA) and related to three thousand pulmonary cancers per year among people who do not smoke in the U.S [14–16]. Among very young children, the incidence of pneumonia, bronchitis, and bronchiolitis is reported to increase two-fold and the effects to be proportional with the frequency of smoking and smokers at the home [17].

The odor of ETS can be eliminated by ventilation, but not meaning that health risks are also removed as it is not possible to totally remove tobacco smoke [18]. The most effective remedy is strict smoking prohibition in the work-place or adapting special smoking rooms with separate ventilation to the outside [19].

### 4.2 Other combustible products causing health problems

**Carbon monoxide** is an odorless and colorless asphyxiant due to carboxyhemoglobin (COHb) resulting from CO binding to Hb, impeding oxygen transport.

**Nitrogen dioxide (NO) and sulfur dioxide ($SO_2$)** particularly irritate ocular, nasal, pharyngeal and respiratory tract mucosa. Acute bronchoconstriction by sulfur dioxide can be observed in asthma cases or as a hypersensitivity reaction. Continued exposure to elevated levels of nitrogen dioxide may result in acute or chronic bronchitis [20].

| Signs and symptoms | Environmental tobacco smoke | Other combustion products | Biological pollutants | Volatile organics | Heavy metals | Sick building syndrome |
|---|---|---|---|---|---|---|
| Respiratory | | | | | | |
| Rhinitis, nasal congestion | Yes | Yes | Yes | Yes | No | Yes |
| Epistaxis | No | No | No | Yes[1] | No | No |
| Pharyngitis, cough | Yes | Yes | Yes | Yes | No | Yes |
| Wheezing, worsening asthma | Yes | Yes | No | Yes | No | Yes |
| Dyspnea | Yes[2] | No | Yes | No | No | Yes |
| Severe lung disease | No | No | No | No | No | Yes[3] |
| Other | | | | | | |
| Conjunctival irritation | Yes | Yes | Yes | Yes | No | Yes |
| Headache or dizziness | Yes | Yes | Yes | Yes | Yes | Yes |
| Lethargy, fatigue, malaise | No | Yes[4] | Yes[5] | Yes | Yes | Yes |
| Nausea, vomiting, anorexia | No | Yes[4] | Yes | Yes | Yes | No |
| Cognitive impairment, personality change | No | Yes[4] | No | Yes | Yes | Yes |
| Rashes | No | No | Yes | Yes | Yes | No |
| Fever, chills | No | No | Yes[6] | No | Yes | No |
| Tachycardia | No | Yes[4] | No | No | Yes | No |
| Retinal hemorrhage | No | Yes[4] | No | No | No | No |
| Myalgia | No | No | No | Yes[5] | No | Yes |
| Hearing loss | No | No | No | Yes | No | No |

[1]*Associated especially with formaldehyde.*
[2]*In asthma.*
[3]*Hypersensitivity pneumonitis, Legionnaires' Disease.*
[4]*Particularly associated with high CO levels.*
[5]*Hypersensitivity pneumonitis, humidifier fever.*
[6]*With marked hypersensitivity reactions and Legionnaires' Disease.*

**Table 2.**
*Diagnostic quick reference.*

**Key Signs/Symptoms;** nausea/emesis, dizziness, headache, fatigue, ocular and upper respiratory tract irritation, tachycardia, chronic cough, confusion wheezing, hypercarboxyhemoglobinemia, increased frequency of angina in cardiovascular patients.

Diagnostic Leads

- Types of heating, cooking or similar equipment and used combustion material (especially charcoal).
- Similar findings/symptoms among households in heating season.

- Odor felt during heating or any damage in the equipment, if they undergo periodic professional inspection.

Remedial Action.

All equipment should be periodically checked by specialized services, especially before each cold season. The ventilation of equipment (including kitchens) is required to be connected to the outdoor environment.

Health Problems Caused By Volatile Organic Compounds (VOCs).

Even at room temperature certain solids or liquids may emit VOCs like formaldehyde, benzene, perchloroethylene for different length of time. They have been observed indoors than outdoors up to 10 times in six locations of the United States as reported by the EPA, even where there were petrochemical plants in use [21].

**Key Signs/Symptoms;** conjunctival irritation, headache, dyspnea, allergic skin reaction, nausea, emesis, fatigue, epistaxis (formaldehyde), nose, throat discomfort, dizziness, declines in serum cholinesterase levels.

Diagnostic Leads

- Presence and quantity of pressed wood products at the resident.

- Exposure to VOCs at work, home, school.

- Exposure to pesticides, paints, or solvents.

Formaldehyde.

Formaldehyde is a possible human carcinogen (EPA). It may irritate ocular (burning or tingling sensations) or respiratory mucosa (dyspnea or wheezing). Formaldehyde vapor may result in hypersensitivity reactions including asthmatics [22].

Pesticides.

They are used in daily life as pesticides and harmful when inhaled or exposed to their vapors or contaminated dusts. Cephalgia, dizziness, muscular weakness, and nausea are the main symptoms. Some of them are considered possible human carcinogens [23].

Remedial Action.

A forced ventilation is required when such products are used. Avoid storage of opened containers of unused paints etc. at home or workplace and similar materials within home or office.

*Health Problems Caused By* Heavy Metals: Airborne Lead And Mercury Vapor.

Key Signs/Symptoms.

Lead Poisoning.

**In Adults;** headache, hearing loss, fatigue, weakness, personality changes, gastrointestinal discomfort/constipation/anorexia/nausea, tremor, coordination loss.

**In Young Children;** abdominal pain, irritability, seizures/loss of consciousness, ataxia, hyperactivity, reduced attention span, (chronic) learning deficits.

**Key Signs/Symptoms of Mercury Poisoning**; headache, tachycardia, muscle cramps or tremors, acrodynia, intermittent fever, neurological dysfunction, personality change.

Diagnostic Leads

- Housing or working in old or restored buildings or nearby busy highway or industrial area.

- Working with lead material (automobile radiators, solder etc.)
- Lead poisoning among people in close contact.
- Exposure to mercury in latex paints or in religious or cultural activities

Remedial Action.

The possible lead dust should be cleaned by wet-mopping. Professional intervention should be sought when handling paints containing lead and adequate protective gear and good-ventilation provided in work areas.

*Health Problems Caused By* SICK BUILDING SYNDROME.

**Key Signs/Symptoms**; headache, dizziness, nausea, sensitivity to odors, lethargy or fatigue, mucosal irritation.

Diagnostic Leads

- Temporal ceasing or aggravation of problems in relation to exposure frequency to suspected building, or seasonality
- Similar complaints in co-workers or peers.

Remedial Action.

The building, HVAC systems or possible conditions should be investigated and examined appropriately.

### 4.3 Health problems caused by two long-term risks: asbestos and radon

Asbestos and radon are among the most publicized indoor air pollutants. Both are known as carcinogens. Their carcinogenic effects are not immediate after prolonged exposure.

Asbestos.

Materials containing asbestos can lose its integrity with time releasing microscopic fibers into the environment. If they remain present in the lungs for many decades as in the case of heavy occupational exposure, they may lead to asbestos-caused pulmonary fibrosis, pulmonary, pleural or peritoneal (including gastrointestinal) carcinoma, or mesothelioma [24].

Radon.

- Radon is a naturally occurring radioactive gas resulting from the decay of radium, itself a decay product of uranium, follows smoking for causing pulmonary malignancies due to the emitted alpha-particles during the decay. It has no odor, color, and taste. Tobacco smoke has a synergistic effect to radon exposure putting smokers and ex-smokers in increased risk.

### 4.4 Health problems caused by animal dander, molds, dust mites, other biologicals

Every home, school, and workplace are subject to biological air pollutants. Some reside outdoor or in human (viruses and bacteria), some in animals or insects (allergens), and some indoor and in water reservoirs (fungi and bacteria), such as humidifiers. High relative humidity is the most important factor contributing to the growth and dissemination of biological agents like house-dust mite populations or fungal growth on damp surfaces. They may cause infections by invading human tissues; hypersensitivity by activating the immune system; and toxicosis by direct effects of toxins [25].

**Key Signs/Symptoms;** rhinitis, dyspnea, cough, chest tightness, recurrent fever, recognized infectious disease, malaise, conjunctival inflammation, exacerbation of asthma.

Diagnostic Leads.

Infectious disease:

- Mounting evidence regarding the workplace, home, etc. as a source place (although very difficult) like presence of a reservoir or disseminator of biologicals
- Evidence of mold growth (visible growth or odors)?

Hypersensitivity disease:

- Relative humidity consistently above 50%.
- Presence of humidifiers or other water-spray systems, proper maintenance.
- History of flooding or leaks or other sources of surface wetting.
- Pets, cockroaches or rodents in the place.

Toxicosis and/or irritation:

- Appropriate ventilation with fresh air.
- Relative humidity consistently above 50% or below 30%.
- Presence of humidifiers or other water-spray systems.
- Evidence of mold growth (visible growth or odors)?
- Presence of bacterial odors (fishy or locker-room smells)?

Remedial Action

- Adequate outdoor air ventilation.
- Cleaning of water reservoirs and chlorination of potable water systems
- Repairing of leaks and seepage.
- Keeping relative humidity below 50%
- Controlling exposure to pets.
- Regular vacuuming of carpets and furniture.

- Covering of mattresses. Washing in hot water (>54.4°C to kill dust mites in soft materials)

Distinguishing whether indoor air pollution originates from the home or workplace.

Some information may help to determine the presence of an indoor air quality problem at workplace:

- Symptoms observed to occur at workplace and to disappear when leaving the workplace, their temporal or locality pattern (day, season or location at work)

- Similar complaint in co-workers.

- Any diagnosis related to IAQ by a physician [26].

## 5. Criteria for workplace indoor air quality

There are some international guiding principles set for indoor air quality. The recommended guidelines define indoor air quality issues with legal standards. These guidelines are prepared and updated by professionals. There is limited information in the World, particularly on concentration guidelines and standards proposed for indoor air pollutants. Only in the United Kingdom and USA, there are concentration guidelines and standards proposed for indoor air pollutants.

- **World Health Organization:** WHO issued various guidelines aiming at protecting the public health from risks arising from some indoor pollutants such as benzene, carbon monoxide, formaldehyde, nitrogen dioxide, polycyclic aromatic hydrocarbon (PAH), benzo [α] pyrene, radon, trichloroethylene, tetrachloroethylene. Reference values offer basic information allowing assessor to decide whether lifelong exposure to these pollutants or exposure to them for a certain approximate period of time impose a significant risk for the health and wellbeing of people [7].

- Committee on the Medical Effects of Air Pollutants (COMEAP) issued "Report on the Impact of Air Pollution on Health for public institutions and agencies. It determined allowed amounts of indoor air pollutants (COMEAP-2004); formaldehyde, benzene, PAHs (as the equivalent of benzo[α]piren), $NO_2$ ve CO for indoors. Air Quality Strategy for England, Wales and Northern Ireland (DETR, 2000; Defra, 2007) sets out policies for the management of indoor air quality. These include air quality targets for ten basic air pollutants for protecting the health of people and the environment, without bearing unacceptable "economic and social costs. These are Particles ($PM_{10}$ and $PM_{2.5}$), $NO_2$, $O_3$, Sulfur dioxide ($SO_2$), PAHs, benzene, 1,3 butadiene, carbon monoxide (CO) and lead. Health and Safety Executive (HSE) supports the regulatory framework for the workplace health and safety in England, Wales and Scotland, in line with the Occupational Health and Safety Law (HSE, 1974).

- Regulation on Control of Substances Hazardous on Health 2002 (HSE, 2002) set out Official Workplace Exposure Limits (WELs) for 500 substances which are listed in the EH40 document (HSE, 2011 and the following revisions), as an action against specific pollutants. These limits include maximum concentrations for short term (15 minutes) and long term (8 hours) exposure in any period of 24 hours.

Although it is mostly related about indoor emissions, exposure limits determine the indoor values which should not be exceeded, no matter what the source is. HSE does not set limits for continuous (24 hours) exposure. For this reason, WELs are not considered as safe concentrations for periods longer than those specified [7].

- A research conducted in Europe in 2005 collected formaldehyde, CO, $NO_2$, benzene and naphthalene under "Group 1: High Risk Chemicals", as they can form in high concentration and impose a significant risk for the health of residents of the building (INDEKS, 2005). "Group 2" included acetaldehyde, toluene, xylene, styrene as the chemicals of second highest risk. These compounds may occur in high concentrations in indoors, but they require less urgent action under risk management practice [7]. Leading institutions regulating the national official rules are American Conference of Governmental Industrial Hygienists (ACGIH) and American Society of Heating and Air-Conditioning Engineers (ASHRAE) [8–10].

The focus point of EPA air quality is to protect the human health against outdoor air. The objective of this Standard is to control emissions of six pollutants during the release of large amounts of vehicle exhaust gas and industrial waste. These standards can be used for the indoor air quality researches as outdoor air quality offer potential contribution to the indoor exposure (**Table 3**) [8–10].

OSHA claims that it has jurisdiction in all workplace environments. These standards are concerned about indoor air quality at office buildings, industrial and construction workplaces. However, OSHA standards have limitations in terms of knowledge of pollutants and limited exposure limits, as OSHA's standards are based on old limits issued by ACGIH in 1968. Original OSA exposure limits were developed out of ACGIH recommendations dated 1968. Up to now, only limits for some chemical pollutants (for example, asbestos and benzene) have been updated. For this reason, general tendency of industrial hygienists to prefer ACGIH Instructions to OSHA limits. Although backed up by federal laws, OSHA limits are rarely exceeded in office environments where one or more pollutant substances are correctly defined. The complicated nature of the indoor air quality is not supported by the OSHA limits [8–10].

| Indoor air pollutants | Permissible concentrations |
|---|---|
| Carbon monoxide (CO) | < 9 ppm |
| Carbon dioxide ($CO_2$) | < 800 ppm |
| Mold | Indoor and outdoor values should be the same |
| Formaldehyde ($CH_2O$) | < 20 $\mu g/m^{3*}$ |
| Total volatile organic compounds (VOC) | < 200 $\mu g/m^{3*}$ |
| 4-Phenyl Cyclohexane (4-PC) | < 3 $\mu g/m^3$ |
| Total particles (PM) | < 20 $\mu g/m^3$ |
| Regular pollutants | < National indoor standard |
| Other pollutants | < 5% of the limit value |

**Above outdoor air concentrations.*

**Table 3.**
*EPA maximum indoor air standards.*

ACGIH is a professional institution which revises and recommends user manuals used for evaluation of Professional workplace exposure by industrial hygienists every year. There are approximately 400 chemicals that are listed with exposure limits of 15 minutes and 8 hours. These directives were prepared to treat the workplace exposure. Professional exposure is generally limited with a period of 8 hour exposure for healthy individuals aged between 18 and 65. For this reason, ACGIH exposure rules do not apply for house exposure for which exposure parameters are different [7–10].

ASHRAE issued a revised mechanical ventilation standard namely "Ventilation for Acceptable Indoor Air Quality Standard" in 1981. ASHRAE developed consensus principles for the indoor air quality in public buildings. The Standard aims at "stating minimum ventilation rates and indoor air quality". Health effects and acceptable exposure limits are based on specific authorized people and their recommendations. For this reason, ASRAE Standard "Ventilation for Acceptable Indoor Air Quality" has become the guideline which is most widely used for the evaluation of indoor air quality in commercial facilities and enterprises. ASHRAE previously issued Standard 62 which is a ventilation standard. This Standard was revised a few times in the following years. The amount of fresh air was specified for smokers and non-smokers separately in 1981. This value was 2,5 L/s foreseen for non-smokers and 10 L/s for smokers, which is four times higher than the one for nonsmokers. Cigarette monopolies prevented the recognition of this Standard by American National Standards Institute (ANSI) and its integration to the building regulations, by conducting intensive propagandas. Application of this Standard will increase the ventilation cost by four times in the buildings where smokers work. In 1989, the acceptable fresh air was accepted to be the amount for which 80% of the people did not express dissatisfaction. The dissatisfaction rate was increased to 20% of the people in the place.

ASHRAE's Standard numbered 1989–2062 introduced the limit of 1000 ppm carbon dioxide for office workers. OHSA in USA defined a limit of 5000 ppm, on the condition that it does not exceed 40 hours a week. ASHRAE's Standard 62 recognized that carbon dioxide is not a pollutant by itself, but it is one of the indicators of air polluted by people. This CO2 amount was 280 ppm before the industrial revolution and it has been continuously increasing due to the combustion of fossil fuel. As a result, global warming caused by greenhouse effect has become a very important public problem. The rate of carbon dioxide in today's air is around 390 ppm and it increases by 2 ppm every year [8–12]. Workplace Exposure Limits (WELs) apply to healthy people who are at working age and directly exposed to pollutants at their workplaces. Generally lower exposure limits are imposed for people who are not healthy or of working age or those who are older than the working age. These lower limits apply even if a person is exposed to a pollutant for a period which is significantly longer than 8 hours or even if the work activities do not directly include pollutant [7–10]. Attention should be paid to the selection of air quality standards applicable to a particular workplace during the selection of the most appropriate air quality standard and guideline. In schools and hospitals which are open to public Access, imposition of the targets set by the World Health Organization's guidelines and the targets introduced by the United Kingdom Air Quality Strategy are more appropriate. As for the industrial environments, Health and Safety Executive (HSE) and Workplace Exposure Limits (WELs) are most appropriate (HSE, 2011) [8, 27]. HSE WELS defines official exposure limits for physically healthy people who are exposed for a nominal period of eight hours a day, five day a week in industrial workplaces. For this reason, the elderly, the young and the disabled who are sensitive to some pollutants should be excluded, when the limits are determined. The people who work in the office environment including the ones who are physically less in form and talented, represent a wider proportion of society, when compared to the workers in the industrial environments. Furthermore, workers may not be aware of the fact that they are exposed to a pollutant, if they are not in contact with it as a direct part of the work they perform, and thus they may not take any measure to protect themselves. Her Majesty's Inspectorate of Pollution (MHIP) recommended that a part of the guidelines given in HSE WELs for the exposure of the general society (1993). In the light of this, it is more appropriate to apply lower outdoor air quality guidelines set by the World Health Organization and United Kingdom Air Quality Strategy (Defra 2007) for exposure of the general population also for the indoor environments. The limits determined by HMIP (1993) may be used for the pollutants which do not fall under this scope [8–12, 28].

## 6. Protection from indoor air pollution at workplaces

An employer should use a systematic approach is needed when treating the air quality at the workplace. The systematic approach to indoor air quality (IAQ) comprise commitment of the management, training, participation of employees, hazard definition, control and program inspection. A management coordinator needs to be assigned for IAQ and a management plan needs to be developed.

## 7. IAQ control methods

There are three main control methods used to decrease the concentration of the indoor air pollutants:

1. Source Management

It includes eliminating the pollutant or replacement of pollutant with a less hazardous one. It is the most effective control method in practice. For example, an employer may install temporary barriers in order to prevent pollutants during construction activity or impose negative pressure on the field in Ref. to the adjacent fields.

2. Engineering Controls

A. **Local exhaust:** Use of local exhaust, such as shading and fume hood are effective in eliminating pollutants which are very concentrated.

B. **General ventilation:** When designed, operated, maintained properly, general ventilation is a measure which control air pollutants of normal amount. A well designed and operating HVAC system ensures comfort, by controlling temperature and relative humidity levels, distributes the amount of air sufficient to meet the needs of ventilation for the building habitants and alleviate and eliminates odors and other pollutants.

C. **Air cleaning:** Firstly, it requires elimination of particles in the air when they pass through HVAC equipment. Generally, HVAC system filtering is used to keep the dirt away from adjustable surfaces during the process of ensuring heat transfer effectiveness.

3. Management Controls

A. **Working Chart:** Managers may significantly decrease the amount of exposure to pollutants in their respective buildings, by using charts. For example, they can take the following actions:

1. Eliminate or decrease the duration in which one worker is exposed to a pollutant (in other words, programming the maintenance or cleaning work in the absence of inhabitants)

2. Decrease the amount of chemical substances used by workers or used near to workers (limit the amount of chemicals used by workers for maintenance or cleaning activities during the activity).

3. Control the place where the chemicals are used (conduct maintenance on moving equipment in a maintenance workshop or place equipment-printers, copy machines to a separate room).

B. **Training:** It is important to give IAQ training to workers. Workers need to informed about the sources and effects of pollutants under their control and smooth operation of ventilation system. Employers can make warnings and/or take measures to decrease personal exposure.

**Cleaning:** Cleaning practices should contain preventing entry of dirt to the environment (using walking doormats), cleaning it when it enters the workplace, discharging the litter, storing the food properly and using minimum amount of cleaning products [7–9, 29].

## 8. Indoor air and coronavirus (COVID-19)

COVID-19 spreads through particles and droplets in the air. Individuals infected by COVID may release particles and droplets of inspiration liquids containing SARS CoV-2 virus to the air (by breathing, talking, singing, exercising, coughing and sneezing). Droplets- particles may continue to disseminate and accumulate indoors of workplaces. Infection may happen in case of inhaling the COVID-19 virus from air in a distance shorter than six feet. The particles from an infected person may move along all room or closed area. The particles may hang in the air for hours even after the person leaves the room. A worker may be exposed to it, if respiration liquids directly jump to the mucosa membrane and if it is sprayed on him or her. The following cases may increase the infection risk:

a. Spending time indoors where the amount of outdoor air and ventilation is poor

b. Performing activities which increases emission of respiration liquids, such as talking loudly, singing, exercising

c. Long term exposure (longer than a few minutes)

d. Spending time in crowded areas (especially without proper mask protection).

Measures to decrease the infection potential of COVID-19:

1. Layout, design of a building, occupancy state, heating, ventilation and acclimatization (HVAC) system may affect the spread potential of virus through air. Although improvements made on ventilation and air cleaning do not alone eliminate of risk of spread for the SARS-CoV-2 virus, EPA recommends that physical distance should be maintained and ventilation should be improved by using outdoor air and air filtering, as the important components of a strategy which includes hygiene and clothing.

2. Cloth masks, face guards or masks should be used. Attention should be paid to surface cleaning, hand washing, disinfection, personal and environment hygiene [8, 30].

## 9. Conclusion

- WHO/ILO Joint Estimates of the Work-related Burden of Disease and Injury (2000–2016: Global Monitoring Report) revealed that approximately 450.000 workers' deaths were associated with air pollution (particle substances, gas, smoke, etc.). This association was reported to be the second most important factor which comes after the factor of working for long hours among the risk factors causing death of workers. Workplace indoor air pollution can significantly increase the health risks of workers, including asthma, allergenic reactions, lung cancer and death as a result of occupational accident [1, 29].

- 1989 EPA Report showed that improved indoor air quality may result in higher productivity and less working day loss. EPA stated that the poor indoor air quality may bring a cost of tens of million dollars to the respective country, employer and the enterprise every year, due to the loss of productivity and medical care cost [8].

- Further research is needed to detect new indoor pollutants which are increasing in number and control their effects. Lifelong awareness, elimination of potential indoor hazards, increased awareness of health service providers and professionals are reported to be important to encourage long term lung health and wellbeing [1].

- Indoor air quality can be defined as an optimal indoor requirement specifying the possible minimum amount of air pollutants to ensure the health, comfort and wellbeing of majority of the workers in any closed workplace, at any given time. Temperature at the workplace depends on relative humidity and flow of air in industry. In addition, indoor air at industrial facilities is associated with the technological processes conducted and contents of the chemicals used. Workplace risk assessment is a means which helps creation of a safer environment and it is a process allowing determination of potential adverse effects imposed on the health of workers. The obligation to determine risk assessments which are both correct and simple led to the development of approaches to asses and control risks, including COSHH (Control of Substances Hazardous to Health) and "Chemical Control Kit" designed to assess chemical risks. Enterprises developed Process Route Healthiness Index (PRHI) to analyze new processes which are not yet in implementation [3].

- It was seen that some people had health symptoms although concentrations of indoor air pollutants are below the indoor air quality guidelines. For this reason, further research is needed to better understand and explain the complicated relationships between IAQ and health symptoms [11].

- A multidisciplinary team comprising experts of occupational medicine, IAQ, building physics and toxicology is recommended for evaluation and management of IAQ problems [5].


## Author details

Ferdi Tanir* and Burak Mete
Department of Public Health, Çukurova University, Adana, Turkey

*Address all correspondence to: ftanir@gmail.com

## References

[1] Seguel JM, Merrill R, Seguel D, Campagna AC. Indoor air quality. American Journal of Lifestyle Medicine. 2016;**11**(4):284-295. DOI: 10.1177/1559827616653343

[2] Cincinelli A, Martellini T. Indoor air quality and health. International Journal of Environmental Research and Public Health. 2017;**14**(11):1286. DOI: 10.3390/ijerph14111286

[3] Cullen MR, Kreiss K. In: Levy BS, Wegman DH, Baron SL, Sokas RK, editors. Indoor Air Quality. Occupational and Environmental Health. Sixth ed. UK: Oxford University Press; 2011. pp. 141-153

[4] Reinhold K, Tint P, Munter R. Indoor air quality in industrial premises. RTU scientific articles Material Science and Applied Chemistry. 2009;**20**:48-57

[5] Carrer P, Wolkoff P. Assessment of indoor air quality problems in office-like environments: Role of occupational health services. International Journal of Environmental Research and Public Health. 2018;**15**:741. DOI: 10.3390/ijerph15040741

[6] Türk Toraks Derneği. İç ortam hava kalitesi. [Internet]. Available from: https://www.toraks.org.tr/site/news/10355 [Accessed: 02.12.2021]

[7] Kukaida V, Upton S. Ensuring Good Indoor Air Quality in Buildings. England: BRE Trust; March 2019

[8] EPA. Indoor Air Quality (IAQ). [Internet]. Available from: https://www.epa.gov/indoor-air-quality-iaq/office-building-occupants-guide-indoor-air-quality [Accessed: 01.12.2021]

[9] OSHA. Indoor Air Quality in Commercial and Institutional Buildings. [Internet]. Available from: https://www.osha.gov/sites/default/files/publications/3430indoor-air-quality-sm.pdf [Access: 05.12.2021]

[10] Bulgurcu H. İç Ortam Hava Kirliliği. Havalandırma ve İç Ortam Hava Kalitesi Notları. [Internet]. Available from: file:///C:/Users/Lenovo/Downloads/Mevzuat%20ve%20standartlar %20(2).pdf [Access: 03.12.2021]

[11] Sakellaris I, Saraga D, Mandin C, de Kluizenaar Y, Fossati S, Spinazzè A, et al. Association of subjective health symptoms with indoor air quality in European office buildings: The OFFICAIR project. Indoor Air. 2021;**2**:426-439. DOI: 10.1111/ina.12749

[12] Al-Zboon KK, Forton OT. Indoor air quality in steel rolling industries and possible health effects. Environment and Natural Resources Journal. 2019;**17**(4):20-29. DOI: 10.32526/ennrj.17.4.2019.28

[13] U.S. Environmental Protection Agency, Office of Air and Radiation. Report to Congress on Indoor Air Quality, Volume II: Assessment and Control of Indoor Air Pollution, pp. I, 4-EPA 400-1-89-001C. 1989

[14] U.S. Department of Health and Human Services. The Health Consequences of Involuntary Smoking, A Report of the Surgeon General. DHHS Publication No. (PHS) 87-8398. 1986

[15] National Research Council. Environmental Tobacco Smoke: Measuring Exposures and Assessing Health Effects. USA: National Academy Press; 1986

[16] National Institute for Occupational Safety and Health. Environmental

Tobacco Smoke in the Workplace: Lung Cancer and Other Health Effects. U.S. Department of Health and Human Services, Current Intelligence Bulletin 54. 1991

[17] U.S. Environmental Protection Agency. Respiratory Health Effects of Passive Smoking: Lung Cancer and Other Disorders

[18] Repace JL, Lowrey AH. An indoor air quality standard for ambient tobacco smoke based on carcinogenic risk. New York State Journal of Medicine. 1985;**85**(7):381-383

[19] American Society of Heating, Refrigeration and Air-conditioning Engineers. Ventilation for Acceptable Air Quality; ASHRAE Standard 62-1989

[20] Samet JM, Marbury MC, Spengler JD. Health effects and sources of indoor air pollution. Part I. Am Rev Respir Dis. 1987;**136**(6):1486-1508. DOI: 10.1164/ajrccm/136.6.1486

[21] U.S. Environmental Protection Agency, Office of Acid Deposition, Environmental Monitoring and Quality Assurance. Project Summary: The Total Exposure Assessment Methodology (TEAM) Study. EPA-600-S6-87-002. 1987

[22] U.S. Environmental Protection Agency, U.S. Public Health Service, and National Environmental Health Association. Introduction to Indoor Air Quality: A Reference Manual, p. 87. EPA-400-3-91-003. 1991

[23] U.S. Environmental Protection Agency Office of Research and Development. Final Report: Nonoccupational Pesticide Exposure Study (NOPES), p. 60. EPA-600-3-90-003. 1990

[24] Rom WN, Travis WD, Brody AR. Cellular and molecular basis of the asbestos-related diseases. The American Review of Respiratory Disease. 1991;**143**(2):408-422. DOI: 10.1164/ajrccm/143.2.408

[25] Burge HA, Feely JC. Indoor air pollution and infectious diseases. In: Samet JM, Spengler JD, editors. Indoor Air Pollution, a Health Perspective. Baltimore MD: Johns Hopkins University Press; 1991. pp. 273-284

[26] EPA. Indoor Air Pollution: An Introduction for Health Professionals [Internet]. Available from: https://www. osha.gov/indoor-air-quality/faqs [Accessed: 05.12.2021]

[27] HSE. Workplace exposure limits. [Internet]. Available from: https://www.hse.gov.uk/coshh/basics/exposurelimits.htm [Accessed: 04.12.2021]

[28] Kosa KH. Indoor Air Quality the Latest Sampling and Analytical Methods. London, New York, USA: CRC press, Taylor & Francis Group; 2011

[29] WHO/ILO joint estimates of the work-related burden of disease and injury, 2000-2016: global monitoring report. [Internet]. Available from: https://www.who.int/publications/i/item/9789240034945 [Accessed: 01.12.2021]

[30] EPA. Indoor Air and Coronavirus (COVID-19). [Internet]. Available from: https://www.epa.gov/ coronavirus/indoor-air-and-coronavirus-covid-19 [Accessed: 05.12.2021]